Elementary Exercises in
Group Theory

For the Dragon Lady

Merrill Mathematics Series
Erwin Kleinfeld, Editor

Elementary Exercises in
Group Theory

F. Lane Hardy
Chicago State College

Charles E. Merrill Publishing Co.
A Bell & Howell Company
Columbus, Ohio

Copyright © 1970 by Charles E. Merrill Publishing Co., Columbus, Ohio. All rights reserved. No part of this book may be reproduced in any form, electronic or mechanical, including photocopy, recording, or any information storage and retrieval system, without permission in writing from the publisher.

Standard Book Number 675-09506-9

Library of Congress Catalog Card Number: 70-91880

1 2 3 4 5 - 74 73 72 71 70

Printed in the United States of America

Preface

The present text was developed for a one-term course in mathematics for liberal arts students. It has long been my opinion that if we wish to furnish this category of students with some conception of modern mathematics, we are more likely to have success by treating one topic in some depth rather than skimming the surface of several. However, the student's usual lack of experience in handling abstract concepts makes the presentation of topics in modern algebra quite difficult. I have found that a partial solution to this problem is the use of programmed exercises. This device can help the student to form the habit of making routine computations which in turn can aid his understanding. As well as the programmed exercises, a large number of problems of the usual type are included.

Very little mathematical background is needed for reading this book; it is assumed only that the student has studied high school algebra. The necessary material in relations, mappings and number theory have been included.

I wish to express my great appreciation to my friend H. E. Hall of DeKalb College for his help; he has checked many of the problems and has written Chapter 2. For an excellent job of typing my thanks go to Mrs. Becky Pattillo. Finally, I wish to thank my wife Frances Hardy for compiling the index.

8 July 1969, Cape Cod, Mass.

To the Student

You will find that each section of this book is followed by a set of programmed exercises; these are usually of a routine nature and quite simple to do. They should be done immediately after you feel you have mastered the material of that section. Be sure to check your answer with that given for each exercise. These exercises are usually followed by typical problems on this material. In some cases solutions are given; in each case you should make a reasonable attempt to solve the problem yourself before reading the solution.

It is highly desirable for you to read other books on the subject as you study this one. The Bibliography contains some appropriate titles; *Groups and Their Graphs* by Grossman and Magnus is especially recommended.

Contents

1 Sets — 1

2 Relations and Mappings — 33

3 Some Number Theory — 83

4 Groups — 129

Appendix — 253

Bibliography — 275

1 Sets

The theory of sets is of fundamental importance in the study of mathematics. For the purposes of this book we require only the language of sets and, in this chapter, we set forth the language of sets which will be useful throughout the remainder of this exposition.

1.1 Notation and Terminology

Our concept of set is that of many things conceived as a single object. For example, we think of Mr. and Mrs. Jones, John, and Jane (their children) as the Jones family. (The term "family" is sometimes used as a synonym for "set.") Also, we think of certain individual advisors to the President collectively and call this group the Cabinet. Such terms as "family," "team," "group," and "duo" suggest the common use of the set concept. We shall use the words "set" and "collection" interchangeably, and the individuals which comprise a set or collection will be called the *members* of the set. In the examples above, Mrs. Jones is a member of the Jones family, and the Postmaster General is a member of the Cabinet.

The customary practice of naming or denoting sets by capital letters and members of sets by lower case letters will be followed here. Usually Latin letters will be em-

ployed: A, B, C, ... will be used for sets and a, b, c, ... for members of sets. As a shorthand notation for the phrase "a is a member of the set A," we use the symbolism

$$a \in A$$

—the symbol "$\in$" standing for the phrase "is a member of." When a is not a member of set A, we write

$$a \notin A$$

A common method of specifying or describing a set is to tabulate its members or elements. The set that we call the Jones family is described by its members: Mr. and Mrs. Jones, John, Jane. Similarly, we specify the President's Cabinet by listing its members: Secretary of State, Secretary of the Treasury, Secretary of Defense, Attorney General, Postmaster General, Secretary of the Interior, Secretary of Agriculture, etc. When this method is used to describe a set, it is common practice to enclose the list of members in braces. As an example of this, we say that the Jones family is

$$\{\text{Mr. Jones, Mrs. Jones, John, Jane}\}$$

Generally, if a, b, c, ... are members, and the only members, of a set A, we write

$$A = \{a, b, c, \ldots\}$$

We read this: "the set A is the set whose members are a, b, c, ..."

Quite often it is not practical to list or tabulate all members of a set. This will always be the case when a large number of objects is involved. In such cases an alternative procedure for specifying the set is to use a property which the members of the set share. Rather than list all colleges in the United States, for example, we describe the set by the property: "is a college in the United States." In other words, we speak of the set whose members are colleges of the United States. Symbolically, we express this using the bracket notation as follows:

$$\{x \mid x \text{ is a college in the United States}\}$$

We read this: "the set of all objects, say x, such that x is a college in the United States." There is obviously nothing special about the use of the letter "x" here, and any other letter of the alphabet would have served just as well. If P stands for any property whatever, we express the fact that an object, say a, has property, P, by writing $P(a)$. For example, P may be the property: "is a whole number." Then $P(10)$ expresses the fact that 10 is a whole number. We can then, for each such property P, describe a set:

$$\{x \mid P(x)\}$$

i.e., the set of *all* objects, x, having the property P. If P is the property "is a whole number," then $\{x \mid P(x)\}$ describes the collection of whole numbers.

Example 1-1: The set whose members are 1, 2, 3 is denoted by "$\{1, 2, 3\}$" and also by "$\{y \mid y$ is one of the first three whole numbers$\}$" and also by "$\{x \mid x$ is 1 or x is 2 or x

is 3}." The statements "$1 \in \{1, 2, 3\}$," "$2 \in \{1, 2, 3\}$," and "$3 \in \{1, 2, 3\}$" are true, while "$4 \in \{1, 2, 3\}$" is false.

Example 1-2: Suppose that $P(x)$ expresses the fact that x is a number, and that $x^2 = 1$. Then "$\{x \mid P(x)\}$" describes the set $\{1, -1\}$.

DEFINITION. The sets A, B are the *same* or *equal* provided their members are the same. This is expressed by "$A = B$."

From our definition of equal sets, we have that $\{1, 2, 3\}$ is the same set as $\{y \mid y$ is one of the first three whole numbers$\}$ and also that

$$\{1, -1\} = \{x \mid x \text{ is a number and } x^2 = 1\}$$

It should be emphasized that *a set and its members are distinct from each other*. This, no doubt, is obvious for sets of more than one member, but is equally true for sets of only one member: The set whose only member is 8; i.e., $\{8\}$ is not the same as 8. Generally, $\{a\} \neq a$ for any object a. Sets of only one element are sometimes called *singletons* or *unit sets*.

Consider the sets $A = \{t \mid t \text{ is a triangle}\}$ and $B = \{r \mid r \text{ is a right triangle}\}$. It is clear that every member of B is in particular a triangle, and hence, a member of A. Another way of saying this is: "if $x \in B$, then $x \in A$." This relation between the sets A, B may be expressed by saying that "B is a subset of A."

DEFINITION. A set B is said to be a *subset* of set A provided every member of B is a member of A.

If B is a subset of A, this is symbolically expressed by writing "$B \subseteq A$." Thus, we have

$B \subseteq A$ if and only if every member of B is a member of A, or
$B \subseteq A$ if and only if $x \in B$ implies $x \in A$

From our definition of equality for sets (they have the same members) and our definition of subset, we conclude that

$$A = B \text{ if and only if } A \subseteq B \text{ and } B \subseteq A \tag{1-1}$$

Example 1-3: Let $P(x)$ express: "x is a triangle of Euclidean geometry and the sum of the interior angles of x is $180°$," and let $Q(y)$ express: "y is a triangle of Euclidean geometry." Then if $A = \{x \mid P(x)\}$ and $B = \{y \mid Q(y)\}$, we have $A = B$. This is easily established by using (1-1) above, and observing that $A \subseteq B$ and $B \subseteq A$. It is clear that every member of A is a member of B so that $A \subseteq B$; and also, since in Euclidean geometry *every* triangle has an angle sum of $180°$, $B \subseteq A$.

Example 1-4: If we insist that every property P defines a set, we are led to the conclusion that there are sets having no elements. Consider, for example, the pro-

perty $P(x)$: "x is a triangle of Euclidean geometry having angle sum less than 180°."
Then the set $\{x|P(x)\}$ has no elements since there are no objects having the property in question. One can think of many properties of this kind. The role of this set which has no elements is analogous to that of the number zero; this analogy will become clearer as we proceed. We call such a set a *null set*, and use the symbol "$\emptyset$" to denote any such set. As an immediate consequence of our definition of subset, we have

$$\emptyset \subseteq X \text{ for all sets } X$$

This is true because there are no elements of $\emptyset$ which are *not* in X. This also shows that there can be only one null set since if $\emptyset_1, \emptyset_2$ are both null sets, then $\emptyset_1 \subseteq \emptyset_2$ and $\emptyset_2 \subseteq \emptyset_1$ from the above, and hence by (1-1) $\emptyset_1 = \emptyset_2$.

The notation "$\{\ \}$" is also used to denote the null set, i.e. $\emptyset = \{\ \}$. This is a naturally suggestive notation and may be used to give a "trick" proof that the null set is a subset of every set. For example, suppose that we wish to show that the null set is a subset of $\{1, 2, 3\}$. Proceed as follows:

The statement "$\{2, 4\} \subseteq \{1, 2, 3\}$" is false and the trouble is caused by the 4 on the left-hand side. Therefore, erase the 4 and we have

$$\{2\ \ \} \subseteq \{1, 2, 3\}$$

which is a true statement. Similarly, $\{4\} \subseteq \{1, 2, 3\}$ is false because of the 4 on the left-hand side. Since the 4 is the only element on the left which causes the statement to be false, we may erase it and obtain a true statement:

$$\{\ \ \} \subseteq \{1, 2, 3\}$$

Exercises

Mark each of the sentences True (T) or False (F) in Exercises 1–26. Here "whole number" refers to the set $\{1, 2, 3, 4, \ldots\}$.

1. _____ T $\quad 4 \in \{1, 2, 4, 7\}$.

2. _____ F $\quad 4 \subseteq \{1, 2, 4, 7\}$.

3. _____ T $\quad 4 \in \{a\,|\,a \text{ is a whole number}\}$.

4. _____ F $\quad 5 \subseteq \{p\,|\,p \text{ is a whole number}\}$.

Sets 5

T	5. _____	$3 \in \{9, 1, 1/2, 3, 11\}$.
F	6. _____	$3 \subseteq \{9, 1, 1/2, 3, 11\}$.
T	7. _____	$\{2\,1/2\} \subseteq \{1/2, 3/2, 5/2\}$.
F	8. _____	$\{2\,1/2, 3\,1/2\} \subseteq \{0.25, 1.3, 2.5, 3.25\}$.
T	9. _____	$\{2.5, 2/3\} \subseteq \{x \mid x \text{ is a fraction}\}$.
T	10. _____	$\{y \mid y \text{ is } 2\} \subseteq \{x \mid x \text{ is an even whole number}\}$.
F	11. _____	$\{s \mid s \text{ is the square of a whole number}\}$ is a subset of $\{r \mid r \text{ is a whole number larger than 2}\}$.
T	12. _____	If N is the set of whole numbers, the statement $3 \in N$ is just another way of saying "3 is a whole number."
F	13. _____	The symbols "$\in$" and "$\subseteq$" may be used interchangeably.
F	14. _____	$\{7\} \in \{8, 5, 7\}$.
T	15. _____	$\{7\} \subseteq \{8, 5, 7\}$.
T	16. _____	$7 \in \{8, 5, 7\}$.
F	17. _____	$7 \subseteq \{8, 5, 7\}$.

18. __F__ $7 \subseteq \{x \mid x$ is a whole number$\}$.

19. __T__ $7 \in \{y \mid y$ is a whole number$\}$.

20. __F__ $\{z \mid z$ is 3 or z is 4$\} = \{3\}$.

21. __T__ $\{z \mid z$ is 3 or z is 4$\} = \{3, 4\}$.

22. __F__ $\{m \mid m$ is an odd whole number$\} = \{1, 13, 5, 19\}$.

23. __T__ $\{17, 21, 5, 101\} \subseteq \{q \mid q$ is an odd whole number$\}$.

24. __T__ $\emptyset \subseteq \{5, 13\}$.

25. __F__ $\{11, 4\} \in \emptyset$.

26. __T__ $\{5, 29, 12\} \subseteq \{5, 29, 12\}$.

27. Use appropriate definitions to justify the truth or falsity of each of the following:
    (a) $\{1, 2, 8\} \subseteq \{x \mid x$ is a whole number$\}$.
    (b) $\{w \mid w$ is a whole number$\} \subseteq \{1, 2, 8\}$.
    (c) $\{1, 2, 8\} = \{y \mid y$ is a whole number$\}$.

28. Let Y be a particular point in a plane, and let
    $A = \{c \mid c$ is a circle of radius 4, 6, or 8, and center at $Y\}$
    $B = \{d \mid d$ is a circle of radius 4, 6, or 8$\}$.
    (a) How many members has set A?
    (b) How many members has set B?
    (c) Which of the statements $A \subseteq B$, $B \subseteq A$, $A = B$ are true?

29. Denote by W the set whose members are letters of the Latin alphabet: $a, b, c,$ ... Let V denote the vowels.
  (a) Which of the statements below are true?
      (i) $V \subseteq W$ (ii) $W \subseteq V$ (iii) $W = V$
  (b) If $Q = \{y \mid y$ is a letter of the word "algebra"$\}$, discuss the following statements:
      (i) $Q \subseteq V$ (ii) $Q \subseteq W$ (iii) $Q = \{r, g, l, b, c, a\}$

30. (a) If $X = \{\emptyset, \{\emptyset\}\}$, discuss the following statements:
      (i) $\{\emptyset\} \in X$ (ii) $\emptyset \in X$ (iii) $\emptyset \subseteq X$ (iv) $\{\emptyset\} \subseteq X$
  (b) If $X = \{\emptyset\}$, discuss the statements in (a).
  (c) With $X = \emptyset$ discuss the statements in (a).

31. Prove that if W is a set, $W \subseteq W$.

32. (a) If $A = \{1, 2\}$, how many subsets does A have?
  (b) If $A = \{1, 2, 3\}$, how many subsets does A have?
  (c) Can you guess the number of subsets of a set of 4 elements? Of 5?

33. From the definition of subset, show that $a \in A$ if, and only if, $\{a\} \subseteq A$.

34. Let A, B, C denote sets.
  (a) Prove that if $A \subseteq B$ and $B \subseteq C$, then $A \subseteq C$.
  (b) Prove that if $A \subseteq B$, $B \subseteq C$ and $C \subseteq A$, then $A = B = C$.

35. Prove that for sets X, Y if $X \subseteq Y$, then $a \notin Y$ implies that $a \notin X$.

1.2 Union, Intersection, Complement

If one starts with sets A, B, there are several ways of describing sets in terms of A and B. The sets so described will sometimes be different from both A and B and thus produce for us new sets. Two such descriptions are embodied in the following definitions.

DEFINITION: Let A, B denote sets.
(a) The *union* of A, B is $\{y \mid y \in A \text{ or } y \in B\}$.
(b) The *intersection* of A, B is $\{q \mid q \in A \text{ and } q \in B\}$.

The word "or" as used in the definition of union is used in its *inclusive* sense; i.e., the statement "$y \in A$ or $y \in B$" is true if and only if either *one* or *both* of the statements "$y \in A$," "$y \in B$" is true. The word "or" will consistently be used in this way.

The notations commonly used for union and intersection are "$\cup$" and "$\cap$" respectively; i.e.,

$$A \cup B = \{t \mid t \in A \text{ or } t \in B\}$$

and

$$A \cap B = \{r \mid r \in A \text{ and } r \in B\}$$

Example 1-5: For the sets $A = \{1, 2, 3\}$, $B = \{3, 4, 5\}$, we have $1 \in (A \cup B)$ since it is true that $1 \in A$ or $1 \in B$. Similarly, $2 \in (A \cup B)$, $3 \in (A \cup B)$, and $5 \in (A \cup B)$. Quite obviously, the statement "$x \in A$ or $x \in B$," is false if $x \notin A$ and $x \notin B$. We conclude from the definition of union that 1, 2, 3, 4, 5 are the members and the only members of $A \cup B$:

$$A \cup B = \{1, 2, 3\} \cup \{3, 4, 5\} = \{1, 2, 3, 4, 5\}$$

Now to consider $A \cap B = \{1, 2, 3\} \cap \{3, 4, 5\}$, we note that the statement "$y \in A$ and $y \in B$" is true if and only if $y = 3$. Hence, from the definition of intersection,

$$A \cap B = \{1, 2, 3\} \cap \{3, 4, 5\} = \{3\}$$

Example 1-6: If $X = \{y \mid y \text{ is an even whole number}\}$
$= \{2, 4, 6, 8, \ldots\}$

and

$$Y = \{w \mid w \text{ is an odd whole number}\} = \{1, 3, 5, 7, \ldots\}$$

then

$$X \cup Y = \{1, 2, 3, 4, 5, \ldots\} = \{z \mid z \text{ is a whole number}\}$$

Also

$$X \cap Y = \varnothing$$

Example 1-7: If W, Z are sets for which the statement "$a \in W$ and $a \in Z$" is false for every object a, then $W \cap Z = \varnothing$ from the definition of intersection. (See Example 1-6.)

Example 1-8: Suppose that "$a \in A$" is a true statement for some set A and object a. Then the statement "$a \in A$ or $a \in B$" is true for every set B. This shows that if $a \in A$, then $a \in (A \cup B)$ by the definition of union. Therefore, from the definition of subset,

$$A \subseteq (A \cup B) \text{ for all sets } A, B.$$

In a similar manner it follows that

$$B \subseteq (A \cup B) \text{ for all sets } A, B.$$

Example 1-9: Under what conditions does the union of two sets A, B not result in a set different from both A and B? For instance, what can we conclude from "$A \cup B = A$"? We can conclude that $B \subseteq A$. This is true because of our definitions of union, equality, and subset. The definition of subset requires that each element of B be a member of A for $B \subseteq A$ to be true. But if $x \in B$, $x \in (A \cup B)$ by Example 1-8, and since we have $A \cup B = A$, the definition of equality of sets requires that $x \in A$. Thus, if $x \in B$, $x \in A$. Hence, if $A \cup B = A$, $B \subseteq A$. These considerations then show that we obtain a set different from both A and B by taking the union $A \cup B$, only if $B \nsubseteq A$ and $A \nsubseteq B$.

DEFINITION. If X, Y are sets, the *difference* of X, Y is

$$\{p \mid p \in X \text{ and } p \notin Y\}.$$

For the difference of X, Y we adopt the notation $X \cap Y$; i.e.,

$$X \cap Y = \{p \mid p \in X \text{ and } p \notin Y\}.$$

(We read $X \cap Y$: "X less Y.")

To illustrate this definition, let Z be the set of whole numbers and let X be the set of even whole numbers. Then the statement "$x \in Z$ and $x \notin X$" is true if and only if x is a whole number and x is not even; i.e., if and only if x is a whole number and x is an odd whole number. We conclude that $Z \cap X$ is the set of odd whole numbers.

It is easy to prove that for all sets X, Y

$$X \cup Y = Y \cup X \quad \text{and} \quad X \cap Y = Y \cap X.$$

This usually is expressed by saying that the union and intersection of sets are *commutative* operations. This property of the union and intersection is not shared by the difference operation, however. By this we mean that it is *not* true that for all sets X, Y

$$X \cap Y = Y \cap X$$

To see this, let $X = \{1, 2, 3\}$ and $Y = \{3, 4, 5\}$. Then by definition of "$\cap$,"

$$X \cap Y = \{1, 2\} \quad \text{and} \quad Y \cap X = \{4, 5\}$$

Therefore, $X \cap Y \neq Y \cap X$.

In particular discussions, the sets under consideration will all be subsets of some set, say U. The set U may be referred to as the *universal* set and, in this situation, the following abbreviation is usually adopted: instead of writing $U \cap A$ we write A'. This is read "A-complement." Since it is understood that $x \in U$, we have

$$A' = \{x \mid x \notin A\}$$

10 Sets

It always will be understood in the following that each set A is a subset of some set U whenever we use the notation A'.

Example 1-10: From the definition of A', $x \notin A$ if and only if $x \in A'$. Hence, we may express the difference $X \ominus A$ as follows:

$$X \ominus A = \{t \mid t \in X \text{ and } t \notin A\}$$
$$= \{t \mid t \in X \text{ and } t \in A'\}$$
$$= X \cap A'$$

Exercises

(A) The complements in these questions are to be considered relative to the universal set $N = \{1, 2, 3, 4, \ldots\}$ of all whole numbers. Complete each of the following.

36. $\{1/2, 3, 5\} \cup \{2, 7\} =$

$\{1/2, 3, 5, 2, 7\}$

37. $\{1/2, 3, 5\} \cap \{2, 7\} =$

$\emptyset$

38. $\{1/2, 3, 5\} \ominus \{2, 7\} =$

$\{1/2, 3, 5\}$

39. $\{2, 7\} \ominus \{1/2, 3, 5\} =$

$\{2, 7\}$

40. $\{2, 4, 6, 8\} \cap \{3, 6, 9\} =$

$\{6\}$

41. $\{3, 6, 9\} \cup \{2, 4, 6, 8\} =$

$\{2, 4, 6, 9, 8, 3\}$

Sets   11

42. $\{3, 6, 9\} \cap \{2, 4, 6, 8\} =$

{3, 9}

43. $\{2, 4, 6, 8\} \cap \{3, 6, 9\} =$

{2, 4, 8}

44. $\{t \mid t = 3n \text{ for } n = 1, 2, 3\} =$

{3, 6, 9}

45. $\{t \mid t \text{ is one of the first four multiples of } 4\} =$

{4, 8, 12, 16}

46. $\{q \mid q = 2n \text{ for } n = 1, 2, 3\} =$

{2, 4, 6}

47. $\{s \mid s = 2m \text{ for } m = 2, 3, 4\} \cap \{p \mid p = 3x \text{ for } x = 1, 2, 3, 4\} =$

{6}

48. $\{a \mid a = 2b \text{ for } b = 1, 2, 3, 4, 5, 6\} \cap \{c \mid c = 3d \text{ for } d = 1, 2, 3, 4, 5, 6\} =$

{6, 12}

49. $\{a \mid a = 2b \text{ where } b \text{ is a whole number}\} \cap \{c \mid c = 3d \text{ where } d \text{ is a whole number}\} =$

{x | x = 6t where t is a whole number}

50. $(\{2/5, 4, -1\} \cup \{2/3, 1, 1/2\}) \cap \{4, 1, 1/2\} =$

{2/5, −1, 2/3}

12   Sets

{y | y is an even whole number}

{7, 8, 9, 10, ...} or {n | n is a whole number greater than 6}

{3, 4, 7, 8, 9, 10, ...} or {n | n is a whole number greater than 6 or n = 3 or n = 4}

{1, 2, 3, 4, 5, 6, 7}

{1, 2, 3, 4}

51. {x | x is an odd whole number}' = _____

52. {1, 2, 3, 4, 5, 6}' = _____

53. {1, 2, 5, 6}' = _____

54. {8, 9, 10, 11, 12, ...}' = _____

55. {m | m is a whole number greater than 4}' = _____

(B) In each of the following draw an appropriate conclusion from the given information. Several conclusions are often possible.

56. Given: $X \subseteq \{1, a\}$
    Conclusion: $X = \varnothing$, $X = \{1\}$, $X = \{a\}$ or $X = \{1, a\}$

57. Given: $A \subseteq B$
    Conclusion: Every element of A is an element of B.

58. Given: $X \subseteq \{1, a\}$ and $\{1, a\} \subseteq X$
    Conclusion: _____

$X = \{1, a\}$

(a) $Y = X \cup Y$
(b) $X \subseteq Y$

59. Given: $Y \cup X \subseteq Y$
    Conclusion: _____

60. Given: $A \subseteq \emptyset$
    Conclusion: _____

$A = \emptyset$

(a) $2 \in X, 3 \in X$ and if $x \neq 2, 3, 4, 5$ then $x \notin X$
(b) $X = \{2, 3\}$ or $X = \{2, 3, 4\}$ or $X = \{2, 3, 5\}$ or $X = \{2, 3, 4, 5\}$

61. Given: $X \cup \{5, 4\} = \{2, 3, 5, 4\}$

    Conclusion: _____

62. Given: $M \cap \{1, 2\} = \{1\}$
    Conclusion: _____

$1 \in M$ and $2 \notin M$

63. Given: $a \in \{1, 2, 3\} \cap \{2, 3, 4\}$
    Conclusion: _____

$a = 2$ or $a = 3$

64. Given: $x \in A \cap B$
    Conclusion: _____

$x \in A$ and $x \in B$

65. Given: $W \subseteq X \cup W$

    Conclusion: _____

No conclusion possible (See Example 1–8)

66. Given: $A \subseteq M$ and $M \subseteq B$
    Conclusion: _____

$A \subseteq B$

67. Given: $\{1, 2, 7\} \cap X = \{7\}$

(a) $1 \in X, 2 \in X, 7 \notin X$
(b) $\{1, 2\} \subseteq X, 7 \notin X$

    Conclusion: _____

14 Sets

Answers (left column):

$Y = \{3, 5, 9\}$ or $Y = \{3, 5, 9, 1\}$ or $Y = \{3, 5, 9, 1, 2\}$

$A \subseteq M$

If $x \in A$, then $x \in B$

$X \cap Y = \emptyset$

$Y = X = \emptyset$

$A \subseteq X \cap Y$

$X \cup Y \subseteq A$

$a \in X$

$\{y\} \subseteq X$

$z \notin W$

---

68. Given: $Y \cap \{1, 2\} = \{3, 5, 9\}$
Conclusion: _____

69. Given: $A \cap M = \emptyset$
Conclusion: _____

70. Given: $A \subseteq B$
Conclusion: _____

71. Given: $X \cap Y = X$
Conclusion: _____

72. Given: $X \cap Y = Y$
Conclusion: _____

73. Given: $A \subseteq X$ and $A \subseteq Y$
Conclusion: _____

74. Given: $X \subseteq A$ and $Y \subseteq A$
Conclusion: _____

75. Given: $\{a\} \subseteq X$
Conclusion: _____

76. Given: $y \in X$
Conclusion: _____

77. Given: $z \in W'$
Conclusion: _____

Sets 15

(a) $x \neq 1, x \neq 2, x \neq 3$
(b) $x \notin \{1, 2\}$ and $x \notin \{2, 3\}$

78. Given: $x \notin \{1, 2\} \cup \{2, 3\}$

    Conclusion: _____

(a) $a \neq 1/2, a \neq 3, a \neq 7, a \neq 5/8$
(b) $a \notin \{1/2, 3, 7\}$ and $a \notin \{5/8\}$

79. Given: $a \notin \{1/2, 3, 7\} \cup \{5/8\}$

    Conclusion: _____

$b \notin X$ and $b \notin Y$

80. Given: $b \notin X \cup Y$
    Conclusion: _____

(a) $a \neq 2$
(b) $a \notin \{1, 2\}$ or $a \notin \{2, 3\}$

81. Given: $a \notin \{1, 2\} \cap \{2, 3\}$
    Conclusion: _____

(a) $c \neq 3$, and $c \neq 7$
(b) $c \notin \{1/2, 3, 7\}$ or $c \notin \{1, 5, 3, 7\}$

82. Given: $c \notin \{1/2, 3, 7\} \cap \{1, 5, 3, 7\}$
    Conclusion: _____

$p \notin A$ or $p \notin B$

83. Given: $p \notin A \cap B$
    Conclusion: _____

$q \in C$ and $q \in B$

84. Given: $q \in C \cap B$
    Conclusion: _____

$y \notin X$ or $y \in D$

85. Given: $y \notin X \cap D$
    Conclusion: _____

(a) $z \notin A'$ and $z \notin C'$
(b) $z \in A$ and $z \in C$
(c) $z \in A \cap C$

86. Given: $z \notin A' \cup C'$

    Conclusion: _____

16 Sets

(a) $w \notin B$ or $w \notin D'$
(b) $w \notin B$ or $w \in D$
(c) $w \in B'$ or $w \in D$
(d) $w \in B' \cup D$

87. Given: $w \notin B \cap D'$

Conclusion: _____

(a) $a \in X'$ and $x \in D$
(b) $a \notin X$ and $x \in D$
(c) $a \notin X$ and $x \notin D'$
(d) $a \notin X \cup D'$
(e) $a \in (X \cup D')'$

88. Given: $a \in X' \cap D$

Conclusion: _____

(a) There is an element a such that $a \in A$ and $a \notin Z$.
(b) There is an element a such that $a \in A \cap Z'$.
(c) $A \cap Z' \neq \emptyset$

89. Given: $A \nsubseteq Z$

Conclusion: _____

1.3 The Algebra of Sets

Let us consider a universal set U and all the subsets of U. The operations $\cup$, $\cap$, and $'$ satisfy the following conditions for all the subsets of U.

Property 1. $A \cup \emptyset = A$, $A \cap \emptyset = \emptyset$ for all $A \subseteq U$

Property 2. $A \cup U = U$, $A \cap U = A$ for all $A \subseteq U$

Property 3. $A \cup A' = U$, $A \cap A' = \emptyset$ for all $A \subseteq U$

Property 4. Commutative Laws:
$A \cup B = B \cup A$, $A \cap B = B \cap A$ for all $A, B \subseteq U$

Property 5. Associative Laws:
$A \cup (B \cup C) = (A \cup B) \cup C$, $A \cap (B \cap C) = (A \cap B) \cap C$
for all $A, B, C \subseteq U$

Property 6. Distributive Laws:
$$A \cup (B \cap C) = (A \cup B) \cap (A \cup C),$$
$$A \cap (B \cup C) = (A \cap B) \cup (A \cap C) \text{ for all } A, B, C \subseteq U$$

Property 7. Indempotent Laws:
$$A \cup A = A, A \cap A = A \qquad \text{for all } A \subseteq U$$

Property 8. $A \subseteq (A \cup B), (A \cap B) \subseteq A \qquad$ for all $A, B \subseteq U$
$B \subseteq (A \cup B), (A \cap B) \subseteq B$

In Section 1.2 the relations $A \subseteq (A \cup B)$ and $B \subseteq (A \cup B)$ were proved. We will now prove one of the distributive laws; this proof is rather typical of proofs concerning the elementary properties of sets.

Proof of $A \cap (B \cup C) = (A \cap B) \cup (A \cap C)$. For convenience we will let $A \cap (B \cup C) = X$ and $(A \cap B) \cup (A \cap C) = Y$. Then it will be shown that (i) $X \subseteq Y$ and (ii) $Y \subseteq X$. It will then follow from (1-1) that $X = Y$ and the proof will be complete.

(i) If $x \in X$, then $x \in A$ and $x \in (B \cup C)$ by definition of intersection. But from the definition of union, $x \in B$ or $x \in C$. Thus, if $x \in X$

$$(x \in A \text{ and } x \in B) \quad \text{or} \quad (x \in A \text{ and } x \in C)$$

i.e.,
$$x \in (A \cap B) \quad \text{or} \quad x \in (A \cap C)$$

from the definition of intersection. Finally, we use the definition of union and conclude that $x \in (A \cap B) \cup (A \cap C)$ or that $x \in Y$. This shows that if $x \in X$, then $x \in Y$ and by the definition of subset it follows that $X \subseteq Y$.

(ii) If $y \in Y$, then $y \in (A \cap B)$ or $y \in (A \cap C)$ by definition of union. Then from the definition of intersection,

$$(y \in A \text{ and } y \in B) \quad \text{or} \quad (y \in A \text{ and } y \in C)$$

i.e.,
$$y \in A \text{ and } (y \in B \quad \text{or} \quad y \in C)$$

Using the definitions of intersection and union, it follows that $y \in A \cap (B \cup C) = X$. Hence, if $y \in Y$, then $y \in X$ and by the definition of subset $Y \subseteq X$.

All the Properties 1–8 may be proved from the appropriate definitions. This is not necessary, however, as a proof of some of them may be based upon the others. For example, suppose the relation

$$(A \cap B) \subseteq A$$

has been proved and also suppose the commutative law

$$A \cap B = B \cap A$$

has been proved. Using these together, we may obtain

$$(A \cap B) \subseteq B$$

In Example 1–9 it was shown that $A \subseteq B$ if $A \cup B = B$. Conversely, we may prove that if $A \subseteq B$, then $A \cup B = B$.

Theorem 1. Let A, B be sets. If $A \subseteq B$, then $A \cup B = B$.

Proof. By Property 8 above
$$B \subseteq (A \cup B)$$
Now we show that $(A \cup B) \subseteq B$. If $x \in (A \cup B)$, then $x \in A$ or $x \in B$ by definition of union. But we are given that $A \subseteq B$ so that in either case ($x \in A$ or $x \in B$) $x \in B$ by the definition of subset. Therefore, if $x \in (A \cup B)$, then $x \in B$ and by definition of subset
$$(A \cup B) \subseteq B$$
Thus we have shown that if $A \subseteq B$ then $B \subseteq (A \cup B)$ and $(A \cup B) \subseteq B$ so that by (1-1)
$$A \cup B = B$$

Theorem 2. Let A, B be sets. Then if $A \subseteq B$,
$$A \cap B = A$$

Proof. (The reader should try to make his own proof; the proof of Theorem 1 may be used as a model.)

Theorem 3. Let A, B, X be sets. If $A \subseteq B$, then
$$(A \cup X) \subseteq (B \cup X)$$

Proof. Since $A \subseteq B$, we have by Theorem 2 that $A = A \cap B$. Then

$A \cup X = (A \cap B) \cup X$ by replacing A by $A \cap B$
$ = X \cup (A \cap B)$ by the commutative law of union
$ = (X \cup A) \cap (X \cup B)$ by one of the distributive laws
$ = (A \cup X) \cap (B \cup X)$ by the commutative law of union.

This establishes that
$$(A \cup X) = (A \cup X) \cap (B \cup X)$$
Then by Property 8
$$(A \cup X) \cap (B \cup X) \subseteq (B \cup X)$$

Exercises

(A) The blanks in Exercises 90–94 should be filled in so as to complete the proofs given.

90. *Sample.* Prove that if A is a set, then $A = (A')'$.

Proof. By the definition of equality for sets, we may prove that $A = (A')'$ if we can show that (1) $A \subseteq (A')'$ and (2) $(A')' \subseteq A$. To show that $A \subseteq (A')'$ we must use the definition of <u>subset</u> and show that if $x \in A$, then <u>$x \in (A')'$</u>. Suppose, then, that $x \in A$. Then by definition of complement <u>$x \notin A'$</u>. Therefore, since $x \notin A'$, we may use the definition of <u>complement</u> again and obtain $x \in (A')'$. This shows that if $x \in A$, then $x \in (A')'$ and hence that $A \subseteq (A')'$. We make a similar argument for (2) and show that if $x \in (A')'$, then $x \in A$. If $x \in (A')'$, then by definition of complement we have that $x \notin A'$. If $x \notin A'$, the definition of <u>complement</u> again gives $x \in A$. Thus, (2) has been proved. Therefore, since (1) and (2) have both been proved, we may conclude that $A = (A')'$.

91. Prove that for sets A, B if $A \subseteq B$, then $B' \subseteq A'$.

Proof. We have to assume that $A \subseteq B$ and show that we may conclude from this that $\underline{B' \subseteq A'}$. To prove that $B' \subseteq A'$ we use the definition of <u>subset</u> and show that if $x \in B'$, then $\underline{x \in A'}$. If $x \in B'$, then by definition of <u>complement</u> $x \notin B$. Now since $x \notin B$ we conclude that $x \notin A$ since we are given that $A \subseteq B$. But if $x \notin A$, then by definition of <u>complement</u> $x \in A'$.

20 Sets

$B' \subseteq A'$	Hence, if $x \in B'$, then $x \in A'$ and this shows that $\underline{B' \subseteq A'}$. It has thus been shown that if $\underline{A \subseteq B}$, then $\underline{B' \subseteq A'}$.
$A \subseteq B, B' \subseteq A'$	
	92. Prove that for sets $X, Y, X \cap Y \subseteq X$.
difference, $X \cap Y$	*Proof.* To show that $X \cap Y \subseteq X$ we use the definition of $\underline{\text{difference}}$ and show that if $x \in \underline{X \cap Y}$, then $x \in X$. If $x \in X \cap Y$, then by the definition of $\underline{\text{difference}}$, $x \in X$ and $x \notin Y$. Hence, if $x \in X \cap Y, x \in \underline{X}$. Therefore, $X \cap Y \subseteq X$.
difference	
X	
	93. Prove that for sets $A, B, (A \cap B) \cap B = \emptyset$.
$\emptyset$	*Proof.* Since $\emptyset$ is a subset of every set, we may show that $(A \cap B) \cap B = \underline{\emptyset}$ by showing that $(A \cap B) \cap B \subseteq \emptyset$. To do this we use the definition of $\underline{\text{subset}}$ and show that if $x \in (A \cap B) \cap B$, then $x \in \emptyset$. If $x \in (A \cap B) \cap B$, we have that $x \in \underline{A \cap B}$ and $x \in \underline{B}$ by definition of $\underline{\text{intersection}}$. But $x \in A \cap B$ implies, by the definition of $\underline{\text{difference}}$ that $x \notin \underline{B}$. Therefore, if $x \in (A \cap B) \cap B$, then $x \in B$ and $x \notin B$. Since there can be no such element x, we conclude that $x \in \emptyset$ if $x \in (A \cap B) \cap B$. Thus, it has been shown that $\underline{(A \cap B) \cap B \subseteq \emptyset}$ and since we already know that $\emptyset \subseteq (A \cap B) \cap B$ it may be concluded by the definition of equality for sets that $(A \cap B) \cap B = \emptyset$.
subset	
$A \cap B, B$	
intersection	
difference	
B	
$(A \cap B) \cap B \subseteq \emptyset$	

94. Prove that for sets A, B if $A \cap B' = \emptyset$, then $A \subseteq B$.

Proof. We must show that $A \subseteq B$ under the assumption that $\underline{A \cap B' = \emptyset}$. To show that $A \subseteq B$ we use the definition of $\underline{\text{subset}}$ and show that if $\underline{x \in A}$ then $\underline{x \in B}$. If we proceed by contradiction $x \in A$ and $x \notin B$ we would have $x \in A$ and $x \in B'$ (by definition of $\underline{\text{complement}}$) and therefore by definition of $\underline{\text{intersection}}$ $x \in A \cap B'$. But we are given that $\underline{A \cap B' = \emptyset}$. Therefore, we cannot have $\underline{x \in A}$ and $\underline{x \in B}$. Hence if $x \in A$, then $\underline{x \in B}$ and this shows that $\underline{A \subseteq B}$.

Side notes:
- $A \cap B' = \emptyset$
- subset
- $x \in A, x \in B$
- complement
- intersection
- $A \cap B' = \emptyset$
- $x \in A, x \notin B$
- $x \in B$,
- $A \subseteq B$

(B) Construct proofs in Exercises 95–98.

95. Prove that for sets X, Y, Z, $(X \cap Y) \cap Z \subseteq X \cap (Y \cup Z)$.

96. Prove that for sets X, Y, Z, $X \cap (Y \cup Z) \subseteq (X \cap Y) \cap Z$.

97. For sets X, Y, Z prove that $X \cap (Y \cup Z) = (X \cap Y) \cap Z$.

98. For sets P, Q, R prove that $P \cap (Q \cap R) \subseteq (P \cap Q) \cup R$.

(C) Supply reasons for the steps in the following proofs.

99. *Sample.* If $A \subseteq B$, then $(A \cap X) \subseteq (B \cap X)$.

Proof.

1. $A \subseteq B$

    Reason: Given

2. $A \cup B = B$

   Reason: Theorem 1 and Step 1

3. $B \cap X = (A \cup B) \cap X$

   Reason: From Step 2

4. $(A \cup B) \cap X = X \cap (A \cup B)$

   Reason: Commutative Law for $\cap$

5. $X \cap (A \cup B) = (X \cap A) \cup (X \cap B)$

   Reason: Distributive Law

6. $(X \cap A) \cup (X \cap B) = (A \cap X) \cup (B \cap X)$

   Reason: Commutative Law for $\cap$

7. Therefore, $B \cap X = (A \cap X) \cup (B \cap X)$

   Reason: Steps 3–6

8. $(A \cap X) \subseteq (A \cap X) \cup (B \cap X)$

   Reason: Property 8 for $\cup$

9. $(A \cap X) \subseteq (B \cap X)$

   Reason: Steps 7 and 8

100. If $A \subseteq X$ and $A \subseteq Y$, then $A \subseteq (X \cap Y)$.

   Proof.

   1. $A \subseteq X$ and $A \subseteq Y$

      Reason: Given

   2. $X = A \cup X$ and $Y = A \cup Y$

      Reason: Step 1 and Theorem 1

   3. $(X \cap Y) = (A \cup X) \cap (A \cup Y)$

      Reason: Step 2

   4. $(A \cup X) \cap (A \cup Y) = A \cup (X \cap Y)$

      Reason: Distributive Law

   5. $(X \cap Y) = A \cup (X \cap Y)$

Steps 3 and 4	Reason: _____
	6. $A \subseteq A \cup (X \cap Y)$
Property 8 for $\cup$	Reason: _____
	7. $A \subseteq (X \cap Y)$
Steps 5 and 6	Reason: _____

101. If $A \cap X = \emptyset$ and $A \cup X = U$, then $X = A'$.

Proof.

	1. $A \cap X = \emptyset$ and $A \cup X = U$
Given	Reason: _____
	2. $A' \cap U = A'$
Property 2 for $\cap$	Reason: _____
	3. $A' \cap U = A' \cap (A \cup X)$
Step 1	Reason: _____
	4. $A' \cap (A \cup X) = (A' \cap A) \cup (A' \cap X)$
Distributive Law	Reason: _____
	5. $(A' \cap A) \cup (A' \cap X) = \emptyset \cup (A' \cap X)$
Property 3 for $\cap$	Reason: _____
	6. $\emptyset \cup (A' \cap X) = (A \cap X) \cup (A' \cap X)$
Step 1	Reason: _____
	7. $(A \cap X) \cup (A' \cap X) = (X \cap A) \cup (X \cap A')$
Commutative Law for $\cap$	Reason: _____
	8. $(X \cap A) \cup (X \cap A') = X \cap (A \cup A')$
Distributive Law	Reason: _____
	9. $X \cap (A \cup A') = X \cap U$
Property 3 for $\cup$	Reason: _____
	10. $X \cap U = X$
Property 2 for $\cap$	Reason: _____

24   Sets

| Steps 2–10 | 11. $A' = X$ |
| | Reason: _____ |

102. $(A')' = A$

Proof.

1. $A' \cap A = A \cap A'$
   Reason: Commutative Law for $\cap$

2. $A \cap A' = \varnothing$
   Reason: Property 3 for $\cap$

3. $A' \cap A = \varnothing$
   Reason: Steps 1 and 2

4. $A' \cup A = A \cup A'$
   Reason: Commutative Law for $\cup$

5. $A \cup A' = U$
   Reason: Property 3 for $\cup$

6. $A' \cup A = U$
   Reason: Steps 4 and 5

7. $A' \cap A = \varnothing$, $A' \cup A = U$
   Reason: Steps 3 and 6

8. $A = (A')'$
   Reason: Exercise 101 and Step 7

103. $A \cup A = A$

Proof.

1. $A' \cup (A \cup A) = (A' \cup A) \cup A$
   Reason: Associative Law for $\cup$

2. $(A' \cup A) \cup A = (A \cup A') \cup A$
   Reason: Commutative Law for $\cup$

3. $(A \cup A') \cup A = U \cup A$

Reason	Statement
Property 3 for $\cup$	Reason: _____
	4. $U \cup A = U$
Property 2 for $\cup$ and commutative law for $\cup$	Reason: _____
	5. $A' \cup (A \cup A) = U$
Steps 1–4	Reason: _____
	6. $A' \cap (A \cup A) = (A' \cap A) \cup (A' \cap A)$
Distributive Law	Reason: _____
	7. $(A' \cap A) \cup (A' \cap A) = \varnothing \cup \varnothing$
Property 3 for $\cap$ and commutative law for $\cap$	Reason: _____
	8. $\varnothing \cup \varnothing = \varnothing$
Property 1 for $\cup$	Reason: _____
	9. $A' \cap (A \cup A) = \varnothing$
Steps 6–9	Reason: _____
	10. $A' \cup (A \cup A) = U$ and $A' \cap (A \cup A) = \varnothing$
Steps 5 and 9	Reason: _____
	11. $A \cup A = (A')'$
Step 10 and Exercise 101	Reason: _____
	12. $(A')' = A$
Exercise 102	Reason: _____
	13. $A \cup A = A$
Steps 11 and 12	Reason: _____

(D) Construct proofs for the statements in Exercises 104–110.

104. If $A \subseteq X$ and $B \subseteq X$, then $(A \cup B) \subseteq X$. (Hint: Use Exercise 100 as a model.)

105. $L = (L \cap M') \cup (L \cap M)$. (Hint: Use a distributive law.)

106. $L = L \cap (L \cup M)$ (Hint: Write $L = L \cup \varnothing$ and use a distributive law.)

107. DeMorgan's Laws:
(a) $(L \cap M)' = L' \cup M'$ (Hint: Show that $(L \cap M) \cap (L' \cup M') = \emptyset$ and $(L \cap M) \cup (L' \cup M') = U$; then apply Exercise 101.)
(b) $(L \cup M)' = L' \cap M'$.

108. $(L \cap M')' = L' \cup M$.

109. $L \cap M = (L' \cup M')'$.

110. $(L \cap M') \cup (M \cap L') = (L \cup M) \cap (L \cap M)'$.

1.4 Geometric Representation of Sets

A very convenient device for picturing sets geometrically is illustrated by the figures below. In a plane we represent a universal set U by the interior of some quad-

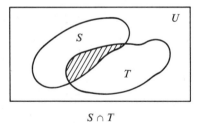

$S \cap T$

Figure 1-1

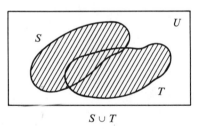

$S \cup T$

Figure 1-2

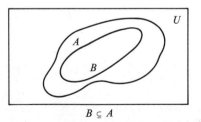

$B \subseteq A$

Figure 1-3

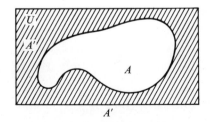

A'

Figure 1-4

rilateral (or other closed figures, for that matter). Then the subsets of U are represented by closed figures interior to U as in Fig. 1-1. Letting the representative figures overlap, we may represent both the union and the intersection of sets by shading appropriate areas. In Fig. 1-1, $S \cap T$ is represented by the shaded area and the shaded area of Fig. 1-2 represents $S \cup T$.

We may represent the relation $B \subseteq A$ as in Fig. 1-3 simply by taking the representative figure for B entirely interior to the figure for A. Consider Example 1-9 of Section 1-2 with the aid of this representation. Other representations are suggested in Figs. 1-5, 1-6, and 1-7.

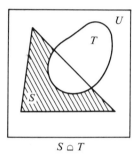

$S \mathbin{\square} T$

Figure 1-5

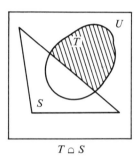

$T \mathbin{\square} S$

Figure 1-6

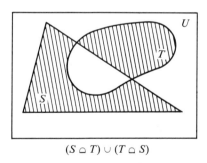

$(S \mathbin{\square} T) \cup (T \mathbin{\square} S)$

Figure 1-7

Example 1-11: To illustrate the validity of Exercise 107(a), we construct separate figures for $(L \cap M)'$ and $L' \cup M'$. In the figures of the left-hand column of Fig. 1-8, we build up $(L \cap M)'$ by stages; in the right-hand column we do a similar thing for $L' \cup M'$. In the two separate columns the final stage is the same shaded area. This, then, suggests the validity of the statement $(L \cap M)' = L' \cup M'$.

28 Sets

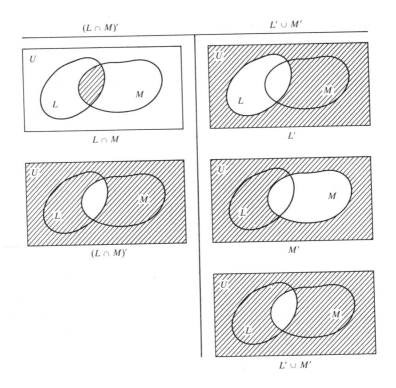

Figure 1-8

Exercises

(A) In the following problems use appropriate set notation to express the shaded areas.

111. *Sample.*

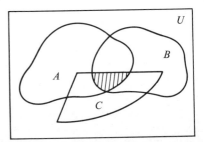

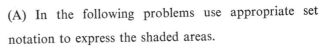

112.

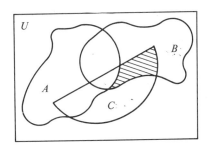

The shaded area may be expressed: _____

C' ∩ B ∩ A'

$(B \cap C) \cap A$

113.

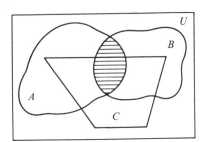

The shaded area may be expressed: _____

$A \cap B$

114.

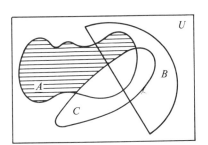

The shaded area may be expressed: _____

$A \cap C$

115.

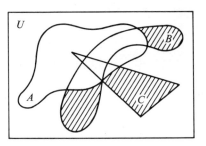

The shaded areas may be expressed: _____

$(B \cap A) \cup (C \cap A)$

116.

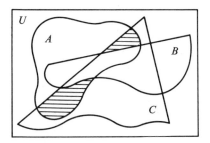

The shaded areas may be expressed: _____

$(A \cap C) \cap B$

117.

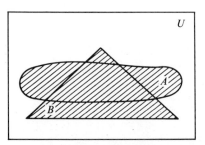

The shaded area may be expressed: _____

$A \cup B$

118.

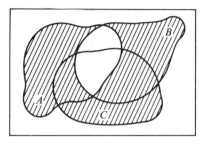

The shaded area may be expressed: _____

$(A \cup B \cup C) \cap (A \cap B)$

119.

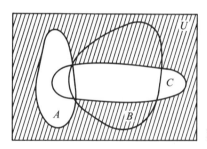

The shaded area may be expressed: _____

$(A \cup C)'$

120.

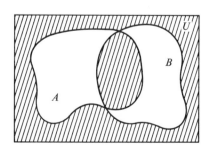

The shaded area may be expressed: _____

$(A \cup B)' \cup (A \cap B)$

(B) Illustrate the validity of each of the following by means of appropriate geometric figures.

121. (a) $L = (L \cap M) \cup (L \cap M)$
    (b) $L = L \cap (L \cup M)$
    (c) $(L \cup M)' = L' \cap M'$
    (d) $(L \cup M')' = L' \cap M$
    (e) $(L \cap M) \cup (M \cap L) = (L \cup M) \cap (L \cap M)$

2 Relations and Mappings

2.1 Cartesian Products

In the previous chapter, the operations union, intersection, and difference of two sets were introduced. We now introduce another method of describing a set in terms of two given ones. In order to define this new concept, the notion of *ordered pair* must be considered.

An ordered pair is first of all a set; however, it is a set with an added property. By our definition of equality for sets, $\{a, b\} = \{b, a\}$. The idea behind the concept of ordered pair is simply that of imposing an order on the set $\{a, b\}$. Consequently, we denote by (a, b), the ordered pair having a as *first element* and b as *second element*. Now we *no longer have the equality* $(a, b) = (b, a)$ *as a general law*. Rather, we have the following rule governing equality of ordered pairs:

$$(a, b) = (c, d) \text{ if and only if } a = c \text{ and } b = d*$$

DEFINITION: If X, Y are sets, the *Cartesian product* of X, Y is

$$\{(a, b) \mid a \in X, b \in Y\}$$

---

*The notion of ordered pair, as we have introduced it, obviously lacks precision. This can be easily remedied by *defining* (a, b) to be the set $\{\{a, b\}, \{b\}\}$. It is then an easy matter to prove the above law of equality. The reader may find the proof of this fact a somewhat interesting exercise.

34 Relations and Mappings

The Cartesian product is commonly denoted by $X \times Y$; i.e.,
$$X \times Y = \{(a, b) \mid a \in X, b \in Y\}$$

Example 2-1: If $X = \{4, 3, 8\}$ and $Y = \{1, a\}$, we have the cartesian products given below:

$$X \times Y = \{(4, a), (4, 1), (3, a), (3, 1), (8, a), (8, 1)\}$$
$$Y \times X = \{(a, 4), (1, 4), (a, 3), (1, 3), (a, 8), (1, 8)\}$$
$$X \times X = \{(4, 4), (4, 3), (4, 8), (3, 4), (3, 3), (3, 8),$$
$$(8, 4), (8, 3), (8, 8)\}$$
$$Y \times Y = \{(a, a), (a, 1), (1, a), (1, 1)\}$$

A useful geometric interpretation of $X \times Y$ may be obtained by representing the members of X and Y on intersecting lines (see Fig. 2-1), say L_1 and L_2 respectively. Then through each point of L_1 which represents an element of X we take a line parallel to L_2; and through each point of L_2 representing a member of Y we draw a line parallel to L_1. The intersections of these various lines represent the members of $X \times Y$ as indicated in Fig. 2-1.

DEFINITION. If X, Y are sets, a subset G of $X \times Y$ is called a *relation* from X to Y. The *domain* of G is the set

$$\mathscr{D}(G) = \{x \mid (x, y) \in G \text{ for some } y \in Y\}$$

and the *range* of G is the set

$$\mathscr{R}(G) = \{z \mid (t, z) \in G \text{ for some } t \in X\}$$

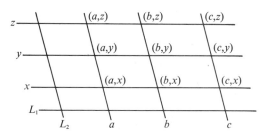

Figure 2-1

It is a rather simple matter to convince oneself that this definition of relation corresponds to our common use of the word. For example, if $X = \{1, 2, 3\}$, then

$$X \times X = \{(1, 1), (1, 2), (1, 3), (2, 1), (2, 2), (2, 3),$$
$$(3, 1), (3, 2), (3, 3)\}$$

Consider now the property "is the same as." This property we ordinarily consider a relation. We may define a set using this property.

$$\{(x, y) \mid x \in X, y \in X, x \text{ is the same as } y\}$$

This is the set $\{(1, 1), (2, 2), (3, 3)\}$, a subset of $X \times X$, and therefore a relation by our definition. The domain as well as the range of this relation is X.

Again, consider the property "is less than." With $X = \{1, 2, 3\}$ as before, define $L = \{(x, y) \mid x \in X, y \in X, x \text{ is less than } y\}$. Then L is the subset $\{(1, 2), (1, 3), (2, 3)\}$ of $X \times X$. One should not conclude from these examples that all relations are as familiar as these; the set $\{(1, 1), (2, 1), (2, 2), (3, 3)\}$ is a relation by our definition, but does not appear to be as familiar as the two examples we have just seen.

If the Cartesian product $X \times Y$ is represented geometrically as suggested above, then a relation R from X to Y may be pictured by marking those points of the grid which belong to R (see Fig. 2-2). It is also convenient to think of a relation as a pairing of elements in its domain with elements in its range. This pairing may be indicated graphically by drawing a line from elements in the domain to the corresponding elements in the range. In Fig. 2-2 the relation $R = \{(1, 1), (2, 2), (3, 3)\}$ has

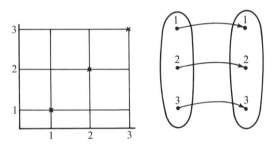

Figure 2-2

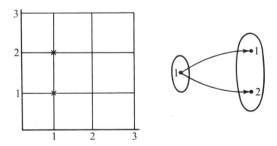

Figure 2-3

been represented in these two ways and in Fig. 2-3 the relation $G = \{(1, 1), (1, 2)\}$ has been represented.

Suppose that X, Y are sets and that the relation R is a subset of $X \times Y$. Then we may associate with R a relation—called the *inverse* of R and denoted by R^{-1}—which is a subset of $Y \times X$. R^{-1} is defined to be the set

$$R^{-1} = \{(w, z) \mid (z, w) \in R\}$$

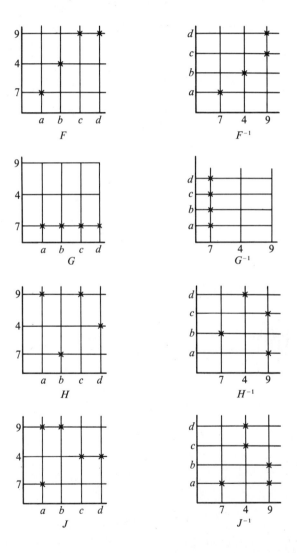

Figure 2-4

In other words, $(w, z) \in R^{-1}$ if and only if $(z, w) \in R$.

Example 2-2: Let F, G, H, J be defined as follows:

$$F = \{(a, 7), (b, 4), (c, 9), (d, 9)\}$$
$$G = \{(a, 7), (b, 7), (c, 7), (d, 7)\}$$
$$H = \{(a, 9), (b, 7), (c, 9), (d, 4)\}$$
$$J = \{(a, 9), (b, 9), (c, 4), (d, 4), (a, 7)\}$$

Then, by the definition of inverse, we have

$$F^{-1} = \{(7, a), (4, b), (9, c), (9, d)\}$$
$$G^{-1} = \{(7, a), (7, b), (7, c), (7, d)\}$$
$$H^{-1} = \{(9, a), (7, b), (9, c), (4, d)\}$$
$$J^{-1} = \{(9, a), (9, b), (4, c), (4, d), (7\ a)\}$$

Figure 2-4 gives geometric representations of these relations.

Exercises

(A) In Problems 1–16, the sets V, W, X are defined as follows: $V = \{1/2, 3, 7\}$, $W = \{8, 3\}$, $X = \{2, 7, 4\}$

Fill in the blanks:

1. $V \times W =$ {(1/2, 8), (3, 8), (7, 8), (1/2, 3), (3, 3), (7, 3)}

2. $V \times X =$ {(1/2, 2), (1/2, 7), (1/2, 4), (3, 2), (3, 7), (3, 4), (7, 2), (7, 7), (7, 4)}

3. $V \times V =$ {(1/2, 1/2), (1/2, 3), (1/2, 7), (3, 1/2), (3, 3), (3, 7), (7, 1/2), (7, 3) (7, 7)}

4. $W \times V =$ {(8, 1/2), (8, 3), (8, 7), (3, 1/2), (3, 3), (3, 7)}

5. $X \times V =$ {(2, 1/2), (2, 3), (2, 7), (7, 1/2), (7, 3), (7, 7), (4, 1/2), (4, 3), (4, 7)}

38   Relations and Mappings

{(8, 8), (8, 3), (3, 8), (3, 3)}

{(2, 2), (2, 7), (2, 4), (7, 2), (7, 7), (7, 4), (4, 4), (4, 2), (4, 7)}

{(2, 8), (2, 3), (7, 8), (7, 3), (4, 8), (4, 3)}

{(8, 1/2), (8, 3), (8, 7), (3, 1/2), (3, 3), (3, 7)}

{(2, 1/2), (7, 1/2), (4, 1/2), (2, 3), (7, 3), (4, 3), (2, 7), (7, 7), (4, 7)}

$V \times V$

$V \times W$

$V \times X$

$W \times W$

$X \times X$

$W \times X$

6. $W \times W =$ _____

7. $X \times X =$ _____
_____

8. $X \times W =$ _____
_____

9. $(V \times W)^{-1} =$ _____
_____

10. $(V \times X)^{-1} =$ _____
_____

11. $(V \times V)^{-1} =$ _____

12. $(W \times V)^{-1} =$ _____

13. $(X \times V)^{-1} =$ _____

14. $(W \times W)^{-1} =$ _____

15. $(X \times X)^{-1} =$ _____

16. $(X \times W)^{-1} =$ _____

(B) In Problems 17–27 the blanks are to be filled in appropriately.

17. *Sample.* If $R = \{(1, 2), (0, 1/3)\}$, then

$\mathscr{D}(R) = \{1, 0\}$ _____

$\mathscr{R}(R) = \{2, 1/3\}$ _____

$R^{-1} = \{(2, 1), (1/3, 0)\}$ _____

18. If $R = \{(10, 7), (a, 5), (a, b)\}$, then

$\mathscr{D}(R) = $ {10, a}

$\mathscr{R}(R) = $ {7, 5, b}

$R^{-1} = $ {(7, 10), (5, a), (b, a)}

19. If $T = \{(14, -1), (3/8, 7/6), (x, 3), (-10, 1)\}$ then

$\mathscr{D}(T) = $ {14, 3/8, x, −10}

$\mathscr{R}(T) = $ {−1, 7/6, 3, 1}

$T^{-1} = $ {(−1, 14), (7/6, 3/8), (3, x), (1, −10)}

20. If $S = \{(x, 2) \mid x \text{ is a whole number}\}$, then

$\mathscr{D}(S) = $ {x | x is a whole number}

$\mathscr{R}(S) = $ {2}

$S^{-1} = $ {(2, x) | x is a whole number}

21. If $Q = \{(t, p) \mid t \text{ is a triangle and } p \text{ is the area of } t\}$,

$\mathscr{D}(Q) = $ {t | t is a triangle}

$\mathscr{R}(Q) = $ {p | p is a positive number}

$Q^{-1} = $ {(p, t) | t is a triangle and p is the area of t}

22. If $L = \{(a, a + 2) \mid a \text{ is a whole number}\}$, then

$\mathscr{D}(L) = $ {a | a is a whole number}

$\mathscr{R}(L) = $ {b | b is a whole number greater than 2}

$L^{-1} = $ {(a + 2, a) | a is a whole number} or {(b, b − 2) | b is a whole number greater than 2}

40   Relations and Mappings

{c | c is a book}

{q | q is a whole number which is the number of pages in some book}

{(q, c) | c is a book and q is the number of pages in c}

{{1, 2, 3}, {1, 2}, {1, 3}, {2, 3}, {1}, {2}, {3}, ∅}

{{2, 8, 1}, {2, 8}, {2, 1}, {1, 8}, {2}, {8}, {1}, ∅}

{{2, 8, 1}, {2, 8}, {2, 8{, {2, 1}, {8, 1}, {2}, {8}, {1}, ∅}

{(B, A) | A ⊆ {2, 8, 1} and B = {2, 8, 1} ⌒ A}

{{0, 1}, {0}, {1}, ∅}

{{(0, 0), (0, 1), (1, 0), (1, 1)}, {(0, 0)}, {(1, 1)}, ∅}

{(D, C) | C ⊆ {0, 1} and D = C × C}

23. If $B = \{(c, q) | c$ is a book and q is the number of pages in $c\}$, then

$\mathscr{D}(B) = $ _____

$\mathscr{R}(B) = $ _____

$B^{-1} = $ _____

24. If $M = \{(A, n) | A \subseteq \{1, 2, 3\}$ and n is the number of elements of $A\}$, then

$\mathscr{D}(M) = $ _____

25. If $N = \{(A, B) | A \subseteq \{2, 8, 1\}$ and $B = \{2, 8, 1\} \cap A\}$, then

$\mathscr{D}(N) = $ _____

$\mathscr{R}(N) = $ _____

$N^{-1} = $ _____

26. If $P = \{(C, D) | C \subseteq \{0, 1\}$ and $D = C \times C\}$, then

$\mathscr{D}(P) = $ _____

$\mathscr{R}(P) = $ _____

$P^{-1} = $ _____

27. If $V = \{(E, t) | E \subseteq \{1, 2, 3, 4\}, E \neq \emptyset$ and t is the smallest element of $E\}$, then

{1, 2, 3, 4}, {1, 3, 4}, {2, 3, 4}, {1, 2, 4},

{3, 1, 2}, {1, 2}, {1, 3}, {1, 4}, {2, 3},

{2, 4}, {3, 4}, {1}, {2}, {3}, {4}}

{1, 2, 3, 4}

$\{(t, E) \mid E \subseteq \{1, 2, 3, 4\}, E \neq \emptyset$ and t is the smallest element of $E\}$

$\mathscr{D}(V) =$ _____

_____

_____

$\mathscr{R}(V) =$ _____

$V^{-1} =$ _____

_____

28. List the members of each of the following sets:
(a) $\{1, 2\} \times \{x\}$ (b) $\{0, 5\} \times \{8, 5\}$

29. Determine the domain and range of each of the following relations:
(a) $\{(3, 2), (2, 2)\}$
(b) $\{(1, 2), (1, 5), (1, 7), (3, 5), (2, 2), (2, 7)\}$
(c) $\{(2, 4), (2, 3), (7, 1), (5, 4)\}$

30. Represent each of the relations in Exercise 29 geometrically in two ways.

2.2 Equivalence Relations

Both the range and domain of the relations considered here will be subsets of the same set, say X. That is, we consider sets of ordered pairs which are subsets of some $X \times Y$ where $X = Y$. If $R \subseteq (X \times X)$, we shall say that R is a *relation on* X.

DEFINITION. Let R be a relation on a set X.
(a) R is *reflexive* on X if and only if $(x, x) \in R$ for all $x \in X$;
(b) R is *symmetric* on X if and only if $(x, y) \in R$ implies $(y, x) \in R$ for all $x, y \in X$;
(c) R is *transitive* on X if and only if $(x, y) \in R$ and $(y, z) \in R$ implies $(x, z) \in R$ for all $x, y, z \in X$;
(d) R is an *equivalence* relation on X if and only if R is reflexive, symmetric, and transitive on X.

Example 2-3: The set $R = \{(1, 1), (2, 2), (2, 3), (3, 3), (3, 1)\}$ is a relation on the set $\{1, 2, 3\}$. R is reflexive since $(1, 1) \in R$, $(2, 2) \in R$ and $(3, 3) \in R$. R is not symmetric since $(2, 3) \in R$ but $(3, 2) \notin R$. R is not transitive since $(2, 3) \in R$, $(3, 1) \in R$ but $(2, 1) \notin R$. R is not an equivalence relation.

Example 2-4: $R = \{(1, 1), (2, 2), (3, 3)\}$ is an equivalence relation on $\{1, 2, 3\}$ as can easily be seen from the definition. The reflexive and symmetric properties are easily seen to hold, and there is nothing to check for the transitive property. What is the common name for this relation?

Example 2-5: $R = \{(1, 2), (2, 1), (1, 1)\}$ is not reflexive on $\{1, 2\}$ since $(2, 2) \notin R$. R is symmetric and transitive. Is R an equivalence relation?

Example 2-6: $R = \{(1, 2), (2, 3), (1, 3)\}$ is not reflexive on $\{1, 2, 3\}$ nor is it symmetric. R is, however, transitive on $\{1, 2, 3\}$.

Notice that by the definition of transitive relation, if we wish to determine whether R is transitive on a set X, we need consider only pairs $(a, b) \in R$ and $(b, c) \in R$. Then R is transitive on X if and only if for each such pair of pairs $(a, c) \in R$. Obviously this is the case if either $a = b$ or $b = c$. This observation will shorten the work in checking some relations for transitivity since we may ignore both the following situations:

$$(a, a) \in R, (a, b) \in R$$

and

$$(a, b) \in R, (b, b) \in R$$

Returning now to graphs of relations, we may ask how the reflexive, symmetric, and transitive properties are reflected in graphs. If R is a relation on the set $X = \{x_1, x_2, x_3, \ldots\}$, we will "picture" R on a frame such as that in Fig. 2-5. (Notice that the angle at A is a right angle, and that points $q \in X$ on AC are the same distance apart as on AD.) The line AB of Fig. 2-5 is called the diagonal. The reflexive property

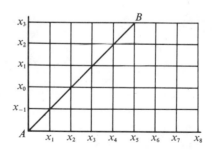

Figure 2-5

is easily interpreted on this frame: A relation R is reflexive on X if and only if each point of the diagonal which is also a point of the grid has an $\times$-mark. For example, Fig. 2-6 (a) is the graph of a nonreflexive relation; Fig. 2-6 (b) shows a reflexive relation.

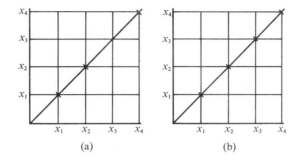

Figure 2-6

Suppose that a relation R is symmetric and that $(a, b) \in R$. Then by definition of "symmetric," we know that $(b, a) \in R$. Now assuming that $a \neq b$, compare the relative positions for the points on the grid corresponding to the pairs (a, b) and (b, a). In Fig. 2-7, these pairs have been plotted. Observe that in Fig. 2-7, if a line is drawn through the point representing (a, b) perpendicular to the line AB, this line passes through the point representing (b, a). The points representing (a, b) and (b, a) are said to be *symmetric* with respect to the line AB.

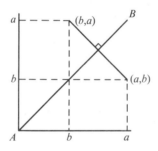

Figure 2-7

On graphs of this type, then, we may interpret a symmetric relation: A relation R is symmetric on X if and only if its graph is symmetric with respect to the diagonal.

In Fig. 2-8, we have the graph of a symmetric relation (a) and a nonsymmetric relation (b). In Fig. 2-8 (b) the point representing (x_3, x_5) is circled—indicating that (x_3, x_5) is not an element of the relation which has been graphed. This fact prevents the relation from being symmetric, since the graph indicates that (x_5, x_3) is a member of the relation.

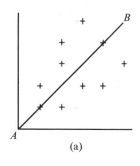

(a)

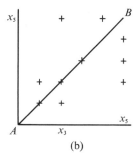

(b)

Figure 2-8

The transitive property is not as easily represented on a graph as the reflexive and symmetric properties and for this reason we will not pursue this point.

If R is a relation, the notation aRb is often used to indicate that $(a, b) \in R$. For example, if $X = \{0, 1, 2, 3, 4, 5\}$, we may define aRb to mean that a is less than b. Then we have

$$R = \{(0, 1), (0, 2), (0, 3), (0, 4), (0, 5), (1, 2), (1, 3),\\ (1, 4), (1, 5), (2, 3), (2, 4), (2, 5), (3, 4), (3, 5),\\ (4, 5)\}$$

which means the same as all of the following:

0R1, 0R2, 0R3, 0R4, 0R5, 1R2, 1R3, 1R4, 1R5, 2R3, 2R4, 2R5, 3R4, 3R5, 4R5

In this notation the definition of equivalence relation may be stated as follows:

A relation R is an equivalence relation on a set X if and only if the following are true:
(a) aRa for all $a \in X$;
(b) aRb implies bRa for all $a, b \in X$;
(c) aRb and bRc imply aRc for all $a, b, c \in X$.

If R is a relation on a set X, a subset $\bar{a}$ of X is associated with each element $a \in X$ as follows:

$$\bar{a} = \{x \mid aRx \text{ and } x \in X\}$$

We may describe the set $\bar{a}$ as all those elements of X that are related by R to a on the right. For the set $X = \{0, 1, 2, 3, 4, 5\}$ and the definition of aRb as above (a is less than b), we have

$$\begin{aligned}
\bar{0} &= \{x \mid 0Rx \text{ and } x \in X\} \\
&= \{x \mid 0 \text{ is less than } x \text{ and } x \in X\} \\
&= \{1, 2, 3, 4, 5\} \\
\bar{1} &= \{x \mid 1Rx \text{ and } x \in X\} \\
&= \{x \mid 1 \text{ is less than } x \text{ and } x \in X\} \\
&= \{2, 3, 4, 5\} \\
\bar{2} &= \{x \mid 2Rx \text{ and } x \in X\} \\
&= \{x \mid 2 \text{ is less than } x \text{ and } x \in X\} \\
&= \{3, 4, 5\} \\
\bar{3} &= \{x \mid 3Rx \text{ and } x \in X\} \\
&= \{x \mid 3 \text{ is less than } x \text{ and } x \in X\} \\
&= \{4, 5\} \\
\bar{4} &= \{x \mid 4Rx \text{ and } x \in X\} \\
&= \{x \mid 4 \text{ is less than } x \text{ and } x \in X\} \\
&= \{5\} \\
\bar{5} &= \{x \mid 5Rx \text{ and } x \in X\} \\
&= \{x \mid 5 \text{ is less than } x \text{ and } x \in X\} \\
&= \emptyset
\end{aligned}$$

Exercises

(A) For Problems 31–45 write in the blanks R, S, T, N according to whether the given relation is reflexive, symmetric, transitive, or none of these.

31. *Sample* If $X = \{1, 3, 5\}$, then the relation $\{(1, 1), (3, 3), (5, 5), (1, 5), (5, 1)\}$ on X is <u>R, S, T</u>.

32. If $X = \{1, 3, 5\}$, then the relation $\{(1, 1), (5, 5), (3, 5), (5, 1)\}$ on X is _____

46   Relations and Mappings

T	33. If $X = \{1, 3, 5\}$, then the relation $\{(5, 1), (1, 3), (5, 3)\}$ on X is _____
R, S, T	34. If N is the set of whole numbers, then the relation $\{(a, b) \mid a \in N \text{ and } b \in N\}$ on N is _____
R, S, T	35. If N is the set of whole numbers, then the relation $\{(a, b) \mid a \in N, b \in N \text{ and } a = b\}$ on N is _____
T	36. N is the set of whole numbers. The relation $\{(x, y) \mid x \text{ is less than } y, x \in N, y \in N\}$ on N is _____
S, T	37. N is the set of whole numbers. The relation $\{(p, q) \mid p \text{ and } q \text{ are both even}, p \in N, q \in N\}$ on N is _____
T	38. $Q = \{1, 2, 3, 4, 5\}$. The relation $\{(a, b) \mid a \in \{1, 2\}, b \in \{4, 5\}\}$ on Q is _____
R, T	39. $Q = \{1, 2, 3, 4, 5\}$. The relation $A \cup B$ on Q, where $A = \{(x, x) \mid x \in Q\}$ and $B = \{(z, w) \mid z \in \{1, 2\}, w \in \{4, 5\}\}$ is _____
S, T	40. $Q = \{1, 2, 3, 4, 5\}$. The relation $\{(x, y) \mid x = y \text{ and } x \in Q \cap \{5\}\}$ on Q is _____
R, S, T	41. $Q = \{1, 2, 3, 4, 5\}$. The relation $\{(x, y) \mid x = y \text{ and } x \in Q \cap \{5\}\}$ on $Q \cap \{5\}$ is _____
S	42. Let X be any set containing at least two elements. Then the relation $\{(a, b) \mid a \in X, b \in X \text{ and } a \neq b\}$ on X is _____

Relations and Mappings   47

S, T

43. Let X be any set containing one element. Then the relation $\{(a, b) \mid a \in X, b \in X \text{ and } a \neq b\}$ on X is _____

R, S, T

44. Let $X = \emptyset$. Then the relation $\{(a, b) \mid a \in X, b \in X \text{ and } a \neq b\}$ on X is _____

R, T

45. Let $Y = \{S \mid S \subseteq N \text{ where } N \text{ is the set of whole numbers}\}$. Then the relation $\{(A, B) \mid A \subseteq B, A \in Y, B \in Y\}$ on Y is _____

(B) In the following exercises supply the necessary information requested.

46. *Sample.* Let $S = \{5, 6, 7, 8\}$ and define aRb, for $a, b \in S$, to mean that $a + b$ is even.

(a) Determine the set R. $R = \underline{\{(5, 5), (6, 6) (7, 7),}$
$\underline{(8, 8), (5, 7), (7, 5), (6, 8), (8, 6)\}}$

(b) Compute $\bar{x}$ for each $x \in S$.

$\bar{5} = \underline{\{5, 7\}}$ $\qquad$ $\bar{6} = \underline{\{6, 8\}}$

$\bar{7} = \underline{\{7, 5\}}$ $\qquad$ $\bar{8} = \underline{\{6, 8\}}$

(c) Determine whether R is reflexive and prove your answer correct.

1st Answer: We can see that R is reflexive since all the pairs $(5, 5), (6, 6), (7, 7), (8, 8)$ are in R.

2nd Answer: For each $a \in S$, $a + a$ is even and hence aRa, showing that R is reflexive by definition.

(d) Determine whether R is symmetric and prove your answer correct.

Answer: If aRb, then $a + b$ is even by definition of

R. Hence, $b + a$ is even and bRa by definition of R. Therefore, R is symmetric by definition of symmetric.

(e) Determine whether R is transitive and prove your answer correct.

Answer: Suppose that aRb and bRc. Then $a + b$ is even and $b + c$ is even. For $x + y$ to be even, where x, y are whole numbers, x, y must both be even or must both be odd. Since $a + b$ is even a, b are both even or both odd. If both are even, then c must also be even since b is even and $b + c$ is even. In this case a, c are both even so that $a + c$ is even and by definition aRc. In a similar fashion, if both a, b are odd, then c is odd. Again $a + c$ is even if both a, c are odd so that aRc. Therefore, if aRb and bRc, then aRc proving that R is transitive.

47. Let $X = \{1, 2, 3, 4, 5, 6\}$ and define aRb to mean that $a = b$ or a is greater than b.

(a) Determine the set R.

$R =$ _____

{(1, 1), (2, 2), (3, 3), (4, 4), (5, 5), (6, 6), (6, 1), (6, 2), (6, 3), (6, 4), (6, 5), (5, 1), (5, 2), (5, 3), (5, 4), (4, 1), (4, 2), (4, 3), (3, 1), (3, 2), (2, 1)}

(b) Compute $\bar{x}$ for each $x \in X$.

$\bar{1} =$ _____

$\bar{2} =$ _____

{1}

{1, 2}

Relations and Mappings 49

{1, 2, 3}
{1, 2, 3, 4}
{1, 2, 3, 4, 5}
{1, 2, 3, 4, 5, 6}

$\bar{3} =$ _____
$\bar{4} =$ _____
$\bar{5} =$ _____
$\bar{6} =$ _____

(c) Is R reflexive? Prove your answer correct.

yes

Answer: _____

$(a, a) \in R$ for all $a \in X$

Proof: _____

(d) Is R symmetric? Prove your answer correct.

Answer: _____

no

$(6, 1) \in R$ but $(1, 6) \notin R$

Proof: _____

(e) Is R transitive? Prove your answer correct.

Answer: _____

yes

If aRb and bRc, then a is greater than or equal to b and b is greater than or equal to c. Hence, a is greater than or equal to c and aRc.

Proof: _____
_____
_____
_____

48. Let W be the set of whole numbers $\{1, 2, 3, 4, 5, 6, \ldots\}$ and define aRb to mean that $b = a + a$.

(a) Determine the set R.

$R =$ _____

$\{(1, 2), (2, 4), (3, 6), (4, 8), (5, 10), \ldots,$
$(a, a + a), \ldots\}$

(b) Compute $\bar{x}$ for each $x \in W$.

$\bar{1} =$ _____ $\bar{2} =$ _____

{2}, {4}
{6}, {8}
{10}, {12}
{x + x}

$\bar{3} =$ _____ $\bar{4} =$ _____
$\bar{5} =$ _____ $\bar{6} =$ _____

Generally, $\bar{x} =$ _____

(c) Is R reflexive? Prove your answer correct.

50 Relations and Mappings

no

$(1, 1) \notin R$ since $1 \neq 1 + 1$

no

$(1, 2) \in R$ but $(2, 1) \notin R$

no

$(1, 2) \in R$ and $(2, 4) \in R$ but $(1, 4) \notin R$

{(2, 2), (5, 5), (7, 7), (3, 3), (6, 6), (12, 12), (8, 8), (10, 10), (2, 5), (2, 7), (5, 2), (7, 2), (5, 7), (7, 5), (3, 6), (6, 3), (6, 12), (12, 6), (3, 12), (12, 3), (8, 10), (10, 8)}

yes

$(a, a) \in R$ for all $a \in X$

yes

If aRb, then a, b are in the same member of $\mathscr{P}$. Hence, b, a are in the same member of $\mathscr{P}$ and bRa.

Answer: _____

Proof: _____

(d) Is R symmetric? Prove your answer correct.

Answer: _____

Proof: _____

(e) Is R transitive? Prove your answer correct.

Answer: _____

Proof: _____

_____

49. Let $X = \{2, 3, 5, 7, 6, 8, 10, 12\}$ and let $\mathscr{P} = \{\{2, 5, 7\}, \{3, 6, 12\}, \{8, 10\}\}$. Define aRb to mean that a and b are in the same member of $\mathscr{P}$. For example, $3R12$ because 3 and 12 are both in $\{3, 6, 12\}$. Determine the set R.

$R = $ _____

_____

_____

_____

_____

(b) Is R reflexive? Prove your answer correct.

Answer: _____

Proof: _____

(c) Is R symmetric? Prove your answer correct.

Answer: _____

Proof: _____

_____

_____

(d) Is R transitive? Prove your answer correct.

yes

If aRb and bRc, then a, b are in the same member of $\mathscr{P}$ and b, c are in the same member of $\mathscr{P}$. Therefore, a, c are in the same member of $\mathscr{P}$ so that aRc.

{2, 5, 7}, {2, 5, 7}
{3, 6, 12}, {8, 10}
{3, 6, 12}, {2, 5, 7}
{8, 10},   {3, 6, 12}

Answer: _____

Proof: _____
_____
_____
_____
_____
_____

(e) Compute $\bar{x}$ for each $x \in X$.

$\bar{2} =$ _____ $\bar{7} =$ _____

$\bar{6} =$ _____ $\bar{10} =$ _____

$\bar{3} =$ _____ $\bar{5} =$ _____

$\bar{8} =$ _____ $\bar{12} =$ _____

2.3 Partitions

Closely related to the concept of "relation" is that of *partition*. In the theorems below the relation that these concepts bear to each other is made precise. The first step towards these theorems is, of course, a definition of "partition."

DEFINITION. Let X be a nonempty set and let $A, B, C, \ldots$ be nonempty subsets of X. Then the set $\{A, B, C, \ldots\}$ is a <u>partition</u> of X if and only if
   (a) X is the union of the sets $A, B, C, \ldots$;
   (b) $Y_1 \cap Y_2 = \varnothing$ for all Y_1, Y_2
such that
$$Y_1 \neq Y_2 \text{ and } Y_1, Y_2 \in \{A, B, C, \ldots\}$$

Example 2.7: Let $X = \{1, 2, 3\}$. The set $\{\{1\}, \{2, 3\}\}$ is a partition of X because (a) $\{1\} \cup \{2, 3\} = X$ and (b) $\{1\} \cap \{2, 3\} = \varnothing$.

Example 2.8: Let $X = \{2, 3, 4, 9\}$. The set $\{\{2, 3\}, \{4\}, \{9, 2\}\}$ is not a partition of X since $\{2, 3\} \cap \{2, 9\} \neq \varnothing$. The set $\{\{2, 3\}, \{4\}, \{9\}\}$ is a partition of X since (a) $\{2, 3\} \cup \{4\} \cup \{9\} = X$ and (b) $\{2, 3\} \cap \{4\} = \varnothing, \{2, 3\} \cap \{9\} = \varnothing, \{4\} \cap \{9\} = \varnothing$.

Example 2.9: The set $\{\{2\}, \{3\}, \{5\}\}$ is not a partition of $\{2, 3, 5, 9\}$ since $\{2\} \cup \{3\} \cup \{5\} \neq \{2, 3, 5, 9\}$.

Example 2.10: If X is the set of all natural numbers $\{1, 2, 3, 4, \ldots\}$ and E, O are the even and odd natural numbers respectively, then E, O is a partition of X since $E \cup O = X$ and $E \cap O = \emptyset$.

Suppose that the set $\{A, B, C, \ldots\}$ is a partition of the set X. Define a relation on X as follows:

aRb means that there is a set Y such that $a \in Y$, $b \in Y$ and $\quad$ (2-1)
$Y \in \{A, B, C, \ldots\}$

We may paraphrase (2–1) by saying that aRb is true if a, b are in the same member of the partition $\{A, B, C, \ldots\}$.

Example 2.11: For the partition $\{\{2, 3\}, \{5, 9\}\}$ of the set $\{2, 3, 5, 9\}$ the set R of (2–1) is

$$\{(2, 2), (3, 3), (2, 3), (3, 2), (5, 5), (9, 9), (5, 9), (9, 5)\}$$

Note here that R is an equivalence relation on the set $\{2, 3, 5, 9\}$.

Theorem 4. If $\{A, B, C, \ldots\}$ is a partition of the set X and a relation R is defined as in (2–1), then R is an equivalence relation on X.

Proof. By the definition of equivalence relation we must show that R is reflexive, symmetric and transitive.
$\quad$(a) *R is reflexive.* Suppose $x \in X$. Then, since X is the union of the sets A, B, $C, \ldots$, there is a $Y \in \{A, B, C, \ldots\}$ such that $x \in Y$. By definition (2–1) it follows that xRx. Therefore, xRx for all $x \in X$ and R is reflexive.
$\quad$(b) *R is symmetric.* Suppose that $x, y \in X$ and that xRy. By (2–1), if xRy, there is a set $Y \in \{A, B, C, \ldots\}$ such that $x \in Y$ and $y \in Y$. But this is the same as saying that there is a set $Y \in \{A, B, C, \ldots\}$ such that $y \in Y$ and $x \in Y$. Then by (2–1) we have yRx, and this shows that R is symmetric.
$\quad$(c) *R is transitive.* Suppose that $x, y, z \in X$, xRy and yRz. Using (2–1), the following conclusions may be obtained:
$\quad$(i) xRy implies that there is a set $Y_1 \in \{A, B, C, \ldots\}$ such that $x \in Y_1$ and $y \in Y_1$;
$\quad$(ii) yRz implies that there is a set $Y_2 \in \{A, B, C, \ldots\}$ such that $y \in Y_2$ and $z \in Y_2$.
From (i) and (ii) we see that $y \in Y_1 \cap Y_2$. Hence, $Y_1 \cap Y_2 \neq \emptyset$ and from the definition of partition we must have $Y_1 = Y_2$. Therefore, we have that $x \in Y_1$ and $z \in Y_1$ and from definition (2–1), xRz. Thus, R is transitive.

$\quad$Theorem 4 tells us how we may use a partition of a set X to obtain an equivalence relation on X. The next result provides a method of producing a partition of a set X from a given equivalence relation. Before proving Theorem 5 let us recall that if R is a relation on a set X, $\bar{a}$ means $\{x \mid aRx, x \in X\}$ where $a \in X$.

Relations and Mappings

Lemma. Let R be an equivalence relation on a nonempty set X. Then for elements $a, b \in X$
  (a) aRb implies $\bar{a} = \bar{b}$;
  (b) $\bar{a} = \bar{b}$ implies aRb.

Proof. (a) To prove that $\bar{a} = \bar{b}$ we will show that $\bar{a} \subseteq \bar{b}$ and $\bar{b} \subseteq \bar{a}$. If $x \in \bar{a}$, then, by definition of $\bar{a}$, aRx. We are given that aRb; hence aRx and aRb and by the symmetric property xRa and aRb. Since R is an equivalence relation, it is transitive and we may conclude that xRb, and by symmetry again bRx. But by the definition of $\bar{b}$, this means that $x \in \bar{b}$. Thus if $x \in \bar{a}$, $x \in \bar{b}$ so that $\bar{a} \subseteq \bar{b}$. If $y \in \bar{b}$, bRy by definition of $\bar{b}$. Then from aRb and bRy it follows that aRy. By definition of $\bar{a}$ it may be concluded that $y \in \bar{a}$. Therefore, $\bar{b} \subseteq \bar{a}$. This concludes the proof that $\bar{a} = \bar{b}$ if aRb. (b) Now suppose that $\bar{a} = \bar{b}$. Since R is an equivalence relation, we have that aRa and bRb. By definition of $\bar{a}, \bar{b}$ we conclude that $a \in \bar{a}$ and $b \in \bar{b}$. But if $\bar{a} = \bar{b}$, this means that $b \in \bar{a}$ and, hence, by definition of $\bar{a}$, aRb.

Theorem 5. If R is an equivalence relation on a nonempty set X, then $\{\bar{a} \mid a \in X\}$ is a partition of X.

Proof. By the definition of partition we must show that $\bar{a} \neq \varnothing$ for each $a \in X$, X is the union of the sets in $\{\bar{a} \mid a \in X\}$ and that $\bar{a} \cap \bar{b} = \varnothing$ if $\bar{a} \neq \bar{b}$. Since R is an equivalence, aRa for each $a \in X$ so that $a \in \bar{a}$. This shows that $\bar{a} \neq \varnothing$. Now let T be the union of the sets in $\{\bar{a} \mid a \in X\}$. We show that $T = X$. Since the set $\bar{a}$ consists of elements of X, it is clear that $T \subseteq X$. Also, if x is any member of X, $x \in \bar{x}$ and hence $x \in T$. Therefore, $X \subseteq T$ and this shows that $T = X$. To prove the last part, we show that if $\bar{a} \cap \bar{b} \neq \varnothing$, then $\bar{a} = \bar{b}$. If $\bar{a} \cap \bar{b} \neq \varnothing$, there is an element $x \in X$ such that $x \in \bar{a}$ and $x \in \bar{b}$. Then aRx and bRx. From the fact that R is an equivalence relation we conclude that aRx and bRx so that aRb. By part (a) of the lemma above, this implies that $\bar{a} = \bar{b}$. This completes the proof.

DEFINITION. If R is an equivalence relation on a set X, the sets $\bar{a}$, for $a \in X$, are called *equivalence classes* and a is called a *representative* of the class $\bar{a}$.

Exercises

T

T

(A) In Exercises 50–59 fill in the blanks with T(true) or F(false). M is the set $\{0, 1, 2, 3, 4, 5, \ldots\}$.

50. _____ The set $\{\{0, 1\}, \{2, 3\}\}$ is a partition of $\{0, 1, 2, 3\}$.

51. _____ The set $\{\{0, 1, 3\}, \{2\}\}$ is a partition of $\{0, 1, 2, 3\}$.

54 Relations and Mappings

T

F

T

T

T

T

F

T

52. _____ The set {{1}, {2}, {3}, {0}} is a partition of the set {0, 1, 2, 3}.

53. _____ The set {{1, 2}, {0, 3}, {2, 3}, {1, 0}} is a partition of the set {0, 1, 2, 3}.

54. _____ The set {{0, 2, 4, 6, ...}, {1, 3, 5, 7, ...}} is a partition of {0, 1, 2, 3, 4, ...}.

55. _____ The set {{$x \mid x$ is an even natural number}, {$y \mid y$ is an odd natural number}} is a partition of the set {$x \mid x$ is a natural number}.

56. _____ The set {{0, 4, 8, 12, ...}, {1, 5, 9, 13, ...}, {2, 6, 10, 14, ...}, {3, 7, 11, 15, ...}} is a partition of M.

57. _____ The set {{$x \mid x = 4r, r \in M$}, {$y \mid y = 4r + 1, r \in M$}, {$z \mid z = 4t + 2, t \in M$}, {$w \mid w = 4u + 3, u \in M$}} is a partition of M.

58. _____ The set {{$x \mid x = 2t, t \in M$}, {$y \mid y = 4s, s \in M$}} is a partition of M.

59. _____ The set {{$x \mid 0Rx$}, {$x \mid 1Rx$}, {$x \mid 2Rx$}, {$x \mid 3Rx$}, ...} is a partition of M if R is an equivalence relation on M.

(B) In Exercises 60–69, supply the information requested.

60. Let $X = \{0, 1, 2, 3\}$. For the partition {{0, 1}, {2},

{(0, 0), (0, 1), (1, 0), (2, 2), (3, 3), (1, 1)}

{(0, 0), (1, 1), (2, 2), (0, 1), (1, 0), (0, 2), (1, 2), (2, 1), (3, 3), (2, 0)}

{(a, a), (b, b), (c, c), (b, c), (c, b), (d, d), (e, e), (d, e), (e, d)}

{(a, a), (b, b), (c, c), (d, d), (e, e)}

{(x, y)| both x, y are even or both x, y are odd}

Relations and Mappings 55

{3}} of X construct the corresponding equivalence relation R as by (2–1).

$R = $ _____

61. Let $X = \{0, 1, 2, 2, 3\}$. For the partition $\{\{0, 1, 2\}, \{3\}\}$ of X construct the corresponding equivalence relation R by (2–1).

$R = $ _____

62. Let $X = \{a, b, c, d, e\}$. For the partition $\{\{a\}, \{b, c\}, \{d, e\}\}$ of X construct the corresponding equivalence relation R by (2–1).

$R = $ _____

63. Let $X = \{a, b, c, d, e\}$. For the partition $\{\{a\}, \{b\}, \{c\}, \{d\}, \{e\}\}$ of X construct the corresponding equivalence relation R by (2–1).

$R = $ _____

64. For the partition $\{\{0, 2, 4, 6, \ldots\}, \{1, 3, 5, 7, \ldots\}\}$ of the set $\{0, 1, 2, 3, 4, 5, \ldots\}$ construct the corresponding equivalence relation R by (2–1).

$R = $ _____

65. Let $T = \{a, b, c, d\}$ and let X be the set of all subsets $\neq \emptyset$ of T. Define xRy, for $x, y \in X$, to

{T, {{a,b,c}, {a,c,d}, {a,b,d}, {b,c,d}},
{{a, b} {a, c}, {a, d}, {b, c}, {b, c}
{b, d}, {c, d}}, {{a}, {b}, {c}, {d}}

mean that x and y have the same number of elements. Is R an equivalence relation on X? Construct the corresponding partition of X.

Answer: _____

66. Let $X = \{3, 5, 7, 8, 9\}$ with the equivalence relation R given as below:

$R = \{(3, 3), (5, 5), (7, 7), (8, 8), (9, 9), (5, 7), (7, 5), (8, 9), (9, 8)\}$. Construct the corresponding partition of X.

Answer: _____

{{3}, {5, 7}, {8, 9}}

67. Let $X = \{3, 5, 7, 8, 9\}$ with the equivalence relation R given as follows:

$R = \{(3, 3), (5, 5), (7, 7), (8, 8), (9, 9), (3, 5), (5, 3), (7, 8), (8, 7), (7, 9), (9, 7), (8, 9), (9, 8)\}$.

Construct the corresponding partition of X.

Answer: _____

{{3, 5}, {7, 8, 9}}

68. Let $X = \{1, 2, 3, 4, 5, \ldots\}$ with the equivalence relation R defined as follows: aRb, for $a, b \in X$, means that $a = b$. Construct the corresponding partition of X.

Answer: _____

{{1}, {2}, {3}, {4}, {5|, ...}

69. Let $X = \{1, 2, 3, 4, 5, \ldots\}$ with the equivalence relation R defined as follows: aRb, for $a, b \in X$,

{{2, 4, 6, 8, 10, ...}, {1, 3, 5, 7, 9, ...}}

means that $a + b$ is even. Construct the corresponding partition of X.

Answer: _____

70. Let L be the set of all lines in a plane and define R as follows: For $a, b \in L$, aRb means that $a = b$ or a is parallel to b. Show that R is an equivalence relation on L and describe the corresponding partition of L.

71. Let C be the set of all circles in a plane and define R as follows: For $a, b \in C$, aRb means that $a = b$ or a and b are concentric. Show that R is an equivalence relation on C and describe the corresponding partition of C.

2.4 Mappings

One of our major concerns here will be the study of certain kinds of relations which we call *mappings* or *functions*. A mapping has the special property that it associates with each element in its domain only one element in its range.

DEFINITION. Let M be a relation from a set X to a set Y. M is a *mapping* of X to Y if and only if the following are true:
  (i) $\mathscr{D}(M) = X$;
  (ii) for all $a \in X$ and $b, c \in Y$ if $(a, b) \in M$ and $(a, c) \in M$, then $b = c$.

Part (ii) of this definition may be expressed by saying that *a mapping M contains no two distinct ordered pairs with the same first element.*

Example 2-12: Let $X = \{1, 2, 3, 4\}$ and $Y = \{2, 3\}$. Then the set $M = \{(1, 2), (2, 2), (3, 3), (4, 3)\}$ is a mapping of X to Y since $\mathscr{D}(M) = X$ and M contains no two distinct ordered pairs with the same first member.

Example 2-13: The set $M = \{(n, 2n) \mid n \text{ is an integer}\}$ is a mapping of the set of integers to the set of integers.

If M is a mapping of X to Y, this is often denoted by writing

$$M: X \longrightarrow Y$$

Also, if $(a, b) \in M$, this is indicated by writing

$$M: a \longrightarrow b$$

or $M(a) = b$. We say that M *maps the set X to the set Y* and that M *maps a to b*. The element b is said to be the *image of a under M*. This notation is suggestive of one of the geometric interpretations we have made of relations. In Fig. 2-9 we have indicated geometrically a mapping M of the set $\{a, b, c, d\}$ to the set $\{*, \square, \triangle\}$.

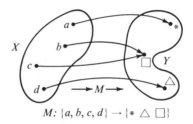

Figure 2-9

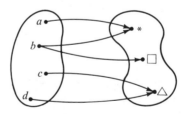

Figure 2-10

Figure 2-10 illustrates a relation which is not a mapping. This fails to be a mapping since the element b has two different images: $*$ and $\square$.

DEFINITION. Let M be a mapping:

$$M: X \longrightarrow Y$$

(a) M is a 1–1 mapping if and only if the following is true: if $a \neq c$ and

$$M: a \longrightarrow b$$
$$M: c \longrightarrow d$$

then $b \neq d$.

(b) M is a mapping of X *onto* Y if and only if the following is true: for each element $y \in Y$, there is an element $x \in X$ such that

$$M: x \longrightarrow y$$

Example 2-14: The mapping

$$M = \{(n, 2n) \mid n \text{ is an integer}\}$$

as a mapping of the integers to the integers is not onto. But it is a mapping of the integers onto the *even* integers. This mapping is 1–1 since if $n \neq m$, then $2n \neq 2m$.

Example 2-15: Let $X = \{1, 2, 3\}$ and $Y = \{4, 5, 6\}$. Define the mapping M as follows:

$$M: 1 \longrightarrow 5$$
$$M: 2 \longrightarrow 4$$
$$M: 3 \longrightarrow 6$$

This is a 1–1 mapping of X onto Y. When finite sets are involved, as is the case here, it is sometimes convenient to denote M as follows:

$$M = \begin{pmatrix} 1 & 2 & 3 \\ 5 & 4 & 6 \end{pmatrix}$$

Notice that the first row above is simply the set X and the images under M are placed directly below; i.e., if $M(a) = b$, then b is placed directly below a.

If M is a mapping of X to Y and $X = Y$, we say that M is a mapping of X to X or that M maps X to itself. The mapping M given below maps the set $\{1, 2, 3\}$ onto itself:

$$M = \begin{pmatrix} 1 & 2 & 3 \\ 3 & 2 & 1 \end{pmatrix}$$

We now wish to consider the total number of 1–1 mappings of a finite set X onto itself. For this purpose we need the following counting principle: *if a task T_1 can be performed in t_1 different ways and a task T_2 can be performed in t_2 different ways, then the tasks T_1, T_2 can be performed successively in $t_1 \cdot t_2$ ways.*

Let us illustrate this counting principle with the following example. Suppose that a motorist wishes to travel from town A to town C by passing through town B, and that there are two roads connecting A to B and three roads connecting B to C. In how many different ways can the motorist make the trip from A to C? (See Fig. 2-11, p. 60.) If the motorist takes Road 1 from A to B, then he may travel from B to C by any one of the three roads connecting these towns; i.e., he may make the trip from A to C in three ways by taking Road 1. Similarly, he may make the trip in three ways if he takes Road 2. We conclude that it is possible to make the trip in a total of six different ways or 2×3 ways. Relating this to the counting principle above we see that the first task T_1 (traveling from A to B) could be performed in 2 ways and the second task T_2 (traveling from B to C) could be performed in 3 ways. Hence, the task of traveling from A to C can be performed in $2 \times 3 = 6$ ways.

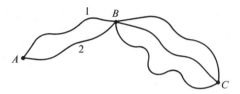

Figure 2-11

Now we apply this counting principle to the problem of finding the total number of 1–1 mappings of a finite set X onto itself. Let us suppose first that $X = \{1, 2, 3\}$. Then any mapping M of X to X may be represented as follows:

$$M = \begin{pmatrix} 1 & 2 & 3 \\ \square & \square & \square \end{pmatrix}$$

The three boxes in row 2 above may be filled with any of the elements 1, 2, 3 and a mapping will result. For example,

$$M = \begin{pmatrix} 1 & 2 & 3 \\ 1 & 3 & 1 \end{pmatrix}$$

is a mapping of X to X (but not 1–1). We may think of filling the first box as a task T_1, filling the second box as a task T_2, and filling the third box as a task T_3. If we do not require that our mappings be 1–1, then each of the tasks T_1, T_2, T_3 may be performed in 3 ways (since X has 3 elements and any one may be used for any one of the boxes). The counting principle applies here and says that T_1, T_2 may be performed successively in $3 \times 3 = 9$ ways. Making another application of this results in the tasks T_1, T_2, T_3 being performed in $9 \times 3 = 27$ ways. Thus, there are 27 mappings of X to X.

Now consider mappings M of $X = \{1, 2, 3\}$ onto X which are 1–1. This time T_1 may be performed in 3 ways as before. But T_2 can now be performed in only two ways because once a choice for the first box has been made, the same element cannot be used for the second box. Similarly, T_3 can be performed in only 1 way. Therefore, T_1, T_2, T_3 can be performed in $3 \times 2 \times 1 = 6$ ways. The total number of 1–1 mappings of 1, 2, 3 onto 1, 2, 3 is 6. We list all of these below:

$$\begin{pmatrix} 1 & 2 & 3 \\ 1 & 2 & 3 \end{pmatrix}, \begin{pmatrix} 1 & 2 & 3 \\ 1 & 3 & 2 \end{pmatrix}, \begin{pmatrix} 1 & 2 & 3 \\ 3 & 2 & 1 \end{pmatrix},$$

$$\begin{pmatrix} 1 & 2 & 3 \\ 2 & 1 & 3 \end{pmatrix}, \begin{pmatrix} 1 & 2 & 3 \\ 2 & 3 & 1 \end{pmatrix}, \begin{pmatrix} 1 & 2 & 3 \\ 3 & 1 & 2 \end{pmatrix}$$

Exercises

(A) For Problems 72–82 write "yes" in the blank if the set satisfies condition (ii) of the definition of mapping; if not, write "no."

72. __No__ $\{(1, 1), (1, 2)\}$

73. __Yes__ $\{(1, 2), (7, 6)\}$

74. __Yes__ $\{(1, 4), (2, 4)\}$

75. __No__ $\{(2, 5), (2, 6)\}$

76. __Yes__ $\{(0, b), (0.5, 7), (9, 0.7)\}$

77. __Yes__ $\{(a, b) \mid b = 2 \text{ and } a \text{ is a whole number}\}$

78. __No__ $\{(a, b) \mid a = 2 \text{ and } b \text{ is a whole number}\}$

79. __No__ $\{(1/2, 1), (2/4, 2), (0.5, 3)\}$

80. __No__ $\{(z^2, z) \mid z \in \{-1, 1\}\}$

81. __Yes__ $\{(n, n + 1) \mid n \text{ is a whole number}\}$

82. __Yes__ $\{(A, B) \mid A \subseteq \{1, 2, 3\} \text{ and } B = \{1, 2, 3\} \cap A\}$

(B) For Problems 83–92, fill in the blanks.

83. *Sample.* If $M = \{(1, 2), (7, 6)\}$, then

62 Relations and Mappings

$M(1) = 2$ _____ $M(7) = 6$ _____

$M: 1 \rightarrow 2$ _____ $M: 7 \rightarrow 6$ _____

84. If $G = \{(1, 4), (2, 4), (5, 6)\}$, then

 4, 4, 6

 $G(1) =$ _____ $G(2) =$ _____ $G(5) =$ _____

 4, 4, 6

 $G: 1 \rightarrow$ _____ $G: 2 \rightarrow$ _____ $G: 5 \rightarrow$ _____

85. If $H = \{(0, b), (0.5, 7), (9, 0.7)\}$, then

 b, 7, 0.7

 $H(0) =$ _____ $H(0.5) =$ _____ $H(9) =$ _____

 b, 7, 0.7

 $H: 0 \rightarrow$ _____ $H: 0.5 \rightarrow$ _____ $H: 9 \rightarrow$ _____

86. If $I = \{(a, b) \mid b = 2$ and a is a whole number$\}$, for each whole number x

 2, 2

 $I(x) =$ _____ $I: x \rightarrow$ _____

87. If $J = \{(q, q^2) \mid q \in \{-1, 1\}\}$, then

 1, 1

 $J(-1) =$ _____ $J(1) =$ _____

 1, 1

 $J: -1 \rightarrow$ _____ $J: 1 \rightarrow$ _____

88. If $K = \{(n, n + 1) \mid n$ is a whole number$\}$, then for any whole number y

 $y + 1$, $y + 1$

 $K(y) =$ _____ $K: y \rightarrow$ _____

89. If $M = \{(A, B) \mid A \subseteq \{1, 2, 3\}$ and $B = \{1, 2, 3\} \cap A\}$,

 $\{1, 2, 3\}$, $\{1, 2, 3\}$

 $M(\varnothing) =$ _____ $M: \varnothing \rightarrow$ _____

 $\{2, 3\}$, $\{2, 3\}$

 $M(\{1\}) =$ _____ $M: \{1\} \rightarrow$ _____

 $\{1, 3\}$, $\{1, 3\}$

 $M(\{2\}) =$ _____ $M: \{2\} \rightarrow$ _____

 $\{1, 2\}$, $\{1, 2\}$

 $M(\{3\}) =$ _____ $M: \{3\} \rightarrow$ _____

 $\{3\}$, $\{3\}$

 $M(\{1, 2\}) =$ _____ $M: \{1, 2\} \rightarrow$ _____

 $\{2\}$, $\{2\}$

 $M(\{1, 3\}) =$ _____ $M: \{1, 3\} \rightarrow$ _____

{1}, {1}
∅, ∅

$M(\{2, 3\}) =$ _____ $M: \{2, 3\} \rightarrow$ _____
$M(\{1, 2, 3\}) =$ _____ $M: \{1, 2, 3\} \rightarrow$ _____

90. If $N = \{(x, y) \mid x$ is a whole number and $y = 0$ if x is even, $y = 1$ if x is odd$\}$, then

0, 0
0, 0
1, 1
1, 1
1, 1
0, 0

$N(2) =$ _____ $N: 2 \rightarrow$ _____
$N(18) =$ _____ $N: 18 \rightarrow$ _____
$N(1) =$ _____ $N: 1 \rightarrow$ _____
$N(99) =$ _____ $N: 9 \rightarrow$ _____
$N(1481) =$ _____ $N: 1481 \rightarrow$ _____
$N(4872) =$ _____ $N: 4872 \rightarrow$ _____

91. If $P = \{(c, d) \mid c \subseteq \{3, 9, 4\}$ and d is the number of elements of $c\}$, then

0, 0
1, 1
1, 1
1, 1
2, 2
2, 2
2, 2
3, 3

$P(\emptyset) =$ _____ $P: \emptyset \rightarrow$ _____
$P(\{3\}) =$ _____ $P: \{3\} \rightarrow$ _____
$P(\{9\}) =$ _____ $P: \{9\} \rightarrow$ _____
$P(\{4\}) =$ _____ $P: \{4\} \rightarrow$ _____
$P(\{3, 9\}) =$ _____ $P: \{3, 9\} \rightarrow$ _____
$P(\{3, 4\}) =$ _____ $P: \{3, 4\} \rightarrow$ _____
$P(\{9, 4\}) =$ _____ $P: \{9, 4\} \rightarrow$ _____
$P(\{3, 9, 4\}) =$ _____ $P: \{3, 9, 4\} \rightarrow$ _____

92. If $Q = \{(X, Y) \mid X \subseteq \{6, 7, 8\}$ and $Y = X \cup \{1, 5, 7, 9\}\}$, then

{1, 5, 7, 9}
{1, 5, 7, 9}
{1, 5, 7, 9, 6}
{1, 5, 7, 9, 6}

$Q(\emptyset) =$ _____
$Q: \emptyset \rightarrow$ _____
$Q(\{6\}) =$ _____
$Q: \{6\} \rightarrow$ _____

64 Relations and Mappings

{1, 5, 7, 9} $Q(\{7\}) =$ _____

{1, 5, 7, 9} $Q: \{7\} \rightarrow$ _____

{1, 5, 7, 9, 8} $Q(\{8\}) =$ _____

{1, 5, 7, 9, 8} $Q: \{8\} \rightarrow$ _____

{1, 5, 7, 9, 6} $Q(\{6, 7\}) =$ _____

{1, 5, 7, 9, 6} $Q: \{6, 7\} \rightarrow$ _____

{1, 5, 7, 9, 6, 8} $Q(\{6, 8\}) =$ _____

{1, 5, 7, 9, 6, 8} $Q: \{6, 8\} \rightarrow$ _____

{1, 5, 7, 9, 8} $Q(\{7, 8\}) =$ _____

{1, 5, 7, 9, 8} $Q: \{7, 8\} \rightarrow$ _____

{1, 5, 7, 9, 6, 8} $Q(\{6, 7, 8\}) =$ _____

{1, 5, 7, 9, 6, 8} $Q: \{6, 7, 8\} \rightarrow$ _____

(C) For Problems 93–102, write "yes" in the blank if the given mapping is 1–1; if it is not 1–1, write "no."

Yes   93. _____ {(1, 2), (7, 6)}

No   94. _____ {(1, 4), (2, 4)}

Yes   95. _____ {(0, 3), (0.5, 7), (9, 0.7)}

No   96. _____ $\{(a, b) \mid b = 2 \text{ and } a \text{ is a whole number}\}$

No   97. _____ $\{(q, q^2) \mid q \in \{-1, 1\}\}$

Yes   98. _____ $\{(n, n + 1) \mid n \text{ is a whole number}\}$

Yes   99. _____ $\{(A, B) \mid A \subseteq \{1, 2, 3\} \text{ and } B = \{1, 2, 3\} \cap A\}$

Relations and Mappings 65

100. _____{(x, y) | x is a whole number and y = 0 if x is even, y = 1 if x is odd}

No

101. _____{(c, d) | c ⊆ {3, 9, 4} and d is the number of elements of c}

No

102. _____{(X, Y) | X ⊆ {6, 7, 8} and Y = X ∪ {1, 5, 7, 9}}

No

(D) Supply the information requested in Problems 103–108.

103. Let $X = \{1\}$. How many 1–1 mappings of X onto X are there?

Answer: _____

1

This mapping is: _____

1 → 1

104. Let $X = \{1, 2\}$. How many 1–1 mappings of X onto X are there?

Answer: _____

2

These mappings are: $\begin{pmatrix} 1 & 2 \\ \_ & \_ \end{pmatrix}$ and $\begin{pmatrix} 1 & 2 \\ \_ & \_ \end{pmatrix}$

1 2, 2 1

105. Let $X = \{1, 2, 3\}$.

(a) How many 1–1 mappings of X onto X are there for which 1 → 1?

Answer: _____

2

(b) How many 1–1 mappings of X onto X are there for which 1 → 2?

2

Answer: _____

66 Relations and Mappings

2

2 × 3

{1, 2, 4}, 6

6

6

(c) How many 1–1 mappings of X onto X are there for which $1 \rightarrow 3$?

Answer: _____

(d) Using (a), (b), (c) above, how many 1–1 mappings of X onto X are there?

Answer: _____

106. Let $X = \{1, 2, 3, 4\}$. In parts (a), (b), (c), and (d) below use the results of Problem 105 and determine the number of 1–1 mappings of X onto X which satisfy the stated condition.

(a) *Sample.* Condition: $4 \rightarrow 4$. The number of 1–1 mappings of X onto X with $4 \rightarrow 4$ is equivalent to the number of 1–1 mappings of $\{1, 2, 3\}$ onto $\{1, 2, 3\}$. Hence there are six 1–1 mappings of X onto X with $4 \rightarrow 4$.

(b) Condition: $4 \rightarrow 3$. The number of 1–1 mappings of X onto X with $4 \rightarrow 3$ is equivalent to the number of 1–1 mappings of $\{1, 2, 3\}$ onto _____. Hence, there are _____ 1–1 mappings of X onto X with $4 \rightarrow 3$.

(c) Condition: $4 \rightarrow 2$.

Answer: _____

(d) Condition: $4 \rightarrow 1$.

Answer: _____

(e) Using (a), (b), (c), (d) above we see that there are six 1–1 mappings of X onto X with $4 \rightarrow 4$; six 1–1 mappings of X onto X with $4 \rightarrow 3$; six 1–1

Relations and Mappings   67

mappings of X onto X with $4 \rightarrow 2$; six 1–1 mappings of X onto X with $4 \rightarrow 1$. Hence, there are a total of _____ 1–1 mappings of $\{1, 2, 3, 4\}$ onto $\{1, 2, 3, 4\}$.

4×6

107. Let $X = \{1, 2, 3, 4, 5\}$. Use the result of 106(e) above to determine the number of 1–1 mappings of X onto X which satisfy the given conditions.

(a) Condition: $5 \rightarrow 5$. Answer: _____   24

(b) Condition: $5 \rightarrow 4$. Answer: _____   24

(c) Condition: $5 \rightarrow 3$. Answer: _____   24

(d) Condition: $5 \rightarrow 2$. Answer: _____   24

(e) Condition: $5 \rightarrow 1$. Answer: _____   24

(f) From parts (a), (b), (c), (d), (e) above we may conclude that there are a total of _____ 1–1 mappings of X onto X.

24×5

108. Using Exercises 103–107, the first 5 places in the table below may be completed. In the top row is the number n of elements in a set X and the corresponding place in the second row is for the number N of 1–1 mappings of X onto X. On the basis of the results obtained for $n = 1, 2, 3, 4, 5$ try to guess the correct number for $n = 6, 7, 8, 9, 10$.

n	1	2	3	4	5
N	1	1·2	1·2·3	1·2·3·4	1·2·3·4·5
			6		7

1·2·3·4·5·6, 1·2·3·4·5·6·7

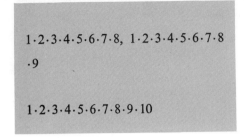 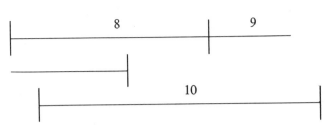

2.5 Products of Mappings and Binary Operations

Let M and N be mappings:
$$M:\ X \longrightarrow Y \quad \text{and} \quad N:\ Y \longrightarrow Z$$

Then if $x \in X$, $M(x) \in Y$ and $N(M(x)) \in Z$. This may be expressed by the following diagram:

$$x \xrightarrow{M} M(x) \xrightarrow{N} N(M(x)) \tag{2-2}$$

Another way to express this is as follows: let $M(x) = y$ and $N(y) = z$. Then
$$M:\ x \longrightarrow y \quad \text{and} \quad N:\ y \longrightarrow z$$

This relation between the mappings M and N is pictured in Fig. 2-12.

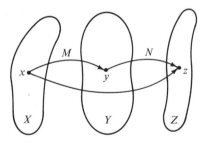

Figure 2-12

As is indicated in Fig. 2-12, the "middle man" may be eliminated to obtain a mapping from X to Z. The mapping from X to Z obtained in this way is called the *product* of M and N and is denoted by MN. From (2-2) and (2-3) we have

$$MN:\ x \longrightarrow N(M(x)) \tag{2-3}$$

and

$$MN:\ x \longrightarrow z \tag{2-4}$$

Example 2-16: Let the mappings M and N be defined as below

$$M = \begin{pmatrix} 1 & 2 & 3 & 4 \\ a & b & c & a \end{pmatrix} \qquad N = \begin{pmatrix} a & b & c & d \\ 7 & 9 & 8 & 6 \end{pmatrix}$$

Then

$$MN = \begin{pmatrix} 1 & 2 & 3 & 4 \\ 7 & 9 & 8 & 7 \end{pmatrix}$$

This may be seen in Fig. 2-13.

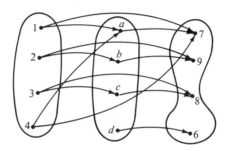

Figure 2-13

Example 2-17: Let X, Y, Z be line segments as shown below. A mapping M from X to Y is defined by taking the image of a point $x \in X$ to be the intersection of the line $\overline{ox}$ with the line segment Y. In the same way a mapping N from Y to Z is defined. The product MN is the mapping of X to Z, as shown in Fig. 2-14.

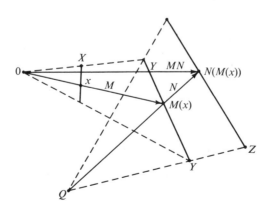

Figure 2-14

Example 2-18: Let X be the set of integers: $\{\ldots, -3, -2, -1, 0, 1, 2, 3, \ldots\}$ and let mappings M, N of X to X be defined as follows:

$$M: \quad x \longrightarrow x + 2 \qquad \text{for each } x \in X$$
$$N: \quad y \longrightarrow 4y - 1 \qquad \text{for each } y \in X$$

Then the product MN is a mapping of X to X defined as follows:

$$MN: \quad x \xrightarrow{M} x + 2 \xrightarrow{N} 4(x+2) - 1$$

or

$$MN: \quad x \longrightarrow 4x + 7 \qquad \text{for each } x \in X$$

DEFINITION. Let X be a set which is not empty. A mapping of $X \times X$ to X is called a *binary operation* on X.

Example 2-19: Let X be the set of integers. Ordinary addition of integers is a binary operation on X. Addition associates with each pair (x, y) of integers an integer $x + y$. This may be denoted by

$$+: \quad (x, y) \longrightarrow x + y \qquad \text{for all } x, y \in X$$

For example,

$$+: \quad (10, 15) \longrightarrow 10 + 15 = 25$$

Example 2-20: Let N be the natural numbers. Subtraction is not a binary operation on N since subtraction does not map $N \times N$ to N, rather it maps $N \times N$ to the set of integers.

Example 2-21: If N is the natural numbers, multiplication is a binary operation on N since every pair of natural numbers may be multiplied to produce a natural number.

If B is a binary operation on a set X, the image of $(x, y) \in X \times X$ will be denoted by xBy, which is in keeping with the way we write images of integers under addition and multiplication.

Let $X = \{1, 2\}$. Then $X \times X = \{(1, 1), (1, 2), (2, 1), (2, 2)\}$ and a binary operation B on X may be defined as follows:

$$B: \quad (1, 1) \longrightarrow 2$$
$$B: \quad (1, 2) \longrightarrow 1$$
$$B: \quad (2, 1) \longrightarrow 1$$
$$B: \quad (2, 2) \longrightarrow 2$$

It is customary to express a binary operation such as B in a "table." The set X is placed in a left column and in the top row as below. Then the images of the various pairs are placed in the appropriate boxes as shown.

B	1	2	X
1	1B1	1B2	
2	2B1	2B2	

X

B	1	2
1	2	1
2	1	2

DEFINITION. Let B be a binary operation on a set X.
(a) B is said to be *commutative* on X if and only if $xBy = yBx$ for all $x \in X$, $y \in X$.
(b) B is said to be *associative* on X if and only if $(xBy)Bz = xB(yBz)$ for all $x, y, z \in X$.

Example 2-22: Addition is both commutative and associative on the set of integers since

$$x + y = y + x \text{ and } x + (y + z) = (x + y) + z$$

for all integers x, y, z.

Example 2-23: Subtraction on the set of integers is not commutative: $2 - 1 = 1$, $1 - 2 = -1$. Therefore, $2 - 1 \neq 1 - 2$. Neither is subtraction associative on the set of integers since

$$3 - (2 - 1) = 2 \text{ and } (3 - 2) - 1 = 0$$

so that $3 - (2 - 1) \neq (3 - 2) - 1$.

Example 2-24: Let the binary operation on the set $\{1, 2\}$ be given by the table below:

B	1	2
1	2	1
2	1	2

The following computations show that B is both commutative and associative:

$1B2 = 2B1 = 1$ and $1B1 = 2B2 = 2$

$(1B1)B1 = 2B1 = 1;$ $\quad 1B(1B1) = 1B2 = 1$
$(1B1)B2 = 2B2 = 2;$ $\quad 1B(1B2) = 1B1 = 2$
$(1B2)B1 = 1B1 = 2;$ $\quad 1B(2B1) = 1B1 = 2$
$(2B1)B1 = 1B1 = 2;$ $\quad 2B(1B1) = 2B2 = 2$
$(1B2)B2 = 1B2 = 1;$ $\quad 1B(2B2) = 1B2 = 1$
$(2B2)B1 = 2B1 = 1;$ $\quad 2B(2B1) = 2B1 = 1$
$(2B1)B2 = 1B2 = 1;$ $\quad 2B(1B2) = 2B1 = 1$
$(2B2)B2 = 2B2 = 2;$ $\quad 2B(2B2) = 2B2 = 2$

Exercises

In Problems 109–114 the products, or images under the products, are to be computed from the given information.

109. *Sample.* If $M: 1 \to 3$ and $N: 3 \to 9$, then
$MN: 1 \to \underline{9}$

110. If $M: a \to 5$ and $N: 5 \to 60$, then
$MN: a \to \underline{}$

[margin: 60]

111. Let Z be the integers with $M: Z \to Z$, $N: Z \to Z$ such that for all $x, y \in Z$, $M: x \to 5x$ and $N: y \to 2y + 1$. Then
$MN: x \xrightarrow{M} \underline{} \xrightarrow{N} \underline{}$
or $MN: x \to \underline{}$
Also, $NM: x \xrightarrow{N} \underline{} \xrightarrow{M} \underline{}$
or $MN: x \to \underline{}$

[margin: $5x$, $2(5x) + 1$
$10x + 1$
$2x + 1$, $5(2x + 1)$
$10x + 5$]

112. Let Z be the integers with $M: Z \to Z$, $N: Z \to Z$ such that for all $x, y \in Z$, $M: x \to 1 - x$ and $N: y \to 5y + 1$. Then
$MN: x \xrightarrow{M} \underline{} \xrightarrow{N} \underline{}$
or $MN: x \to \underline{}$
Also, $NM: x \xrightarrow{N} \underline{} \xrightarrow{M} \underline{}$
or $NM: x \to \underline{}$

[margin: $1 - x$, $5(1 - x) + 1$
$6 - 5x$
$5x + 1$, $1 - (5x + 1)$
$-5x$]

113. Let $X = \{1, 2, 3, 4\}$ and let M, N be the mappings of X to X given below:

Relations and Mappings

$$M = \begin{pmatrix} 1 & 2 & 3 & 4 \\ 3 & 4 & 2 & 1 \end{pmatrix}, \quad N = \begin{pmatrix} 1 & 2 & 3 & 4 \\ 4 & 2 & 1 & 2 \end{pmatrix}$$

Then

$MN: 1 \xrightarrow[M]{} 3 \xrightarrow[N]{} 1; \quad MN: 1 \rightarrow 1$

$MN: 2 \xrightarrow[M]{} \underline{} \xrightarrow[N]{} 2; \quad MN: 2 \rightarrow \underline{}$

$MN: 3 \xrightarrow[M]{} \underline{} \xrightarrow[N]{} 2; \quad MN: 3 \rightarrow \underline{}$

$MN: 4 \xrightarrow[M]{} \underline{} \xrightarrow[N]{} 4; \quad MN: 4 \rightarrow \underline{}$

Therefore, MN may be written:

$$MN = \begin{pmatrix} 1 & 2 & 3 & 4 \\ \underline{} \end{pmatrix}$$

$NM: 1 \xrightarrow[N]{} \underline{} \xrightarrow[M]{} 1; \quad NM: 1 \rightarrow \underline{}$

$NM: 2 \xrightarrow[N]{} \underline{} \xrightarrow[M]{} \underline{}; \quad NM: 2 \rightarrow \underline{}$

$NM: 3 \xrightarrow[N]{} \underline{} \xrightarrow[M]{} \underline{}; \quad NM: 3 \rightarrow \underline{}$

$NM: 4 \xrightarrow[N]{} \underline{} \xrightarrow[M]{} \underline{}; \quad NM: 4 \rightarrow \underline{}$

Therefore,

$$NM = \begin{pmatrix} 1 & 2 & 3 & 4 \\ \underline{} \end{pmatrix}$$

114. Let $X = \{1, 2, 3, 4\}$ and let M, N be the mappings of X to X given below:

$$M = \begin{pmatrix} 1 & 2 & 3 & 4 \\ 2 & 3 & 4 & 1 \end{pmatrix}, \quad N = \begin{pmatrix} 1 & 2 & 3 & 4 \\ 4 & 3 & 2 & 1 \end{pmatrix}$$

Then

$MM: 1 \xrightarrow[M]{} \underline{} \xrightarrow[M]{} \underline{}; \quad MM: 1 \rightarrow \underline{}$

$MM: 2 \xrightarrow[M]{} \underline{} \xrightarrow[M]{} \underline{}; \quad MM: 2 \rightarrow \underline{}$

$MM: 3 \xrightarrow[M]{} \underline{} \xrightarrow[M]{} \underline{}; \quad MM: 3 \rightarrow \underline{}$

74   Relations and Mappings

1, 2, 2

3 4 1 2

2, 3, 3

3, 2, 2

4, 1, 1

1, 4, 4

3 2 1 4

4, 1, 1

3, 4, 4

2, 3, 3

1, 2, 2

1 4 3 2

4, 1, 1

3, 2, 2

2, 3, 3

1, 4, 4

1 2 3 4

2

$MM: 4 \xrightarrow{M} \underline{\quad} \xrightarrow{M} \underline{\quad}$; $MM: 4 \to \underline{\quad\quad}$

$MM = \begin{pmatrix} 1 & 2 & 3 & 4 \\ \underline{\quad\quad\quad} \end{pmatrix}$

$MN: 1 \xrightarrow{M} \underline{\quad} \xrightarrow{N} \underline{\quad}$; $MN: 1 \to \underline{\quad\quad}$

$MN: 2 \xrightarrow{M} \underline{\quad} \xrightarrow{N} \underline{\quad}$; $MN: 2 \to \underline{\quad\quad}$

$MN: 3 \xrightarrow{M} \underline{\quad} \xrightarrow{N} \underline{\quad}$; $MN: 3 \to \underline{\quad\quad}$

$MN: 4 \xrightarrow{M} \underline{\quad} \xrightarrow{N} \underline{\quad}$; $MN: 4 \to \underline{\quad\quad}$

$MN = \begin{pmatrix} 1 & 2 & 3 & 4 \\ \underline{\quad\quad\quad} \end{pmatrix}$

$NM: 1 \xrightarrow{N} \underline{\quad} \xrightarrow{M} \underline{\quad}$; $NM: 1 \to \underline{\quad\quad}$

$NM: 2 \xrightarrow{N} \underline{\quad} \xrightarrow{M} \underline{\quad}$; $NM: 2 \to \underline{\quad\quad}$

$NM: 3 \xrightarrow{N} \underline{\quad} \xrightarrow{M} \underline{\quad}$; $NM: 3 \to \underline{\quad\quad}$

$NM: 4 \xrightarrow{N} \underline{\quad} \xrightarrow{M} \underline{\quad}$; $NM: 4 \to \underline{\quad\quad}$

$NM = \begin{pmatrix} 1 & 2 & 3 & 4 \\ \underline{\quad\quad\quad} \end{pmatrix}$

$NN: 1 \xrightarrow{N} \underline{\quad} \xrightarrow{N} \underline{\quad}$; $NN: 1 \to \underline{\quad\quad}$

$NN: 2 \xrightarrow{N} \underline{\quad} \xrightarrow{N} \underline{\quad}$; $NN: 2 \to \underline{\quad\quad}$

$NN: 3 \xrightarrow{N} \underline{\quad} \xrightarrow{N} \underline{\quad}$; $NN: 3 \to \underline{\quad\quad}$

$NN: 4 \xrightarrow{N} \underline{\quad} \xrightarrow{N} \underline{\quad}$; $NN: 4 \to \underline{\quad\quad}$

$NN = \begin{pmatrix} 1 & 2 & 3 & 4 \\ \underline{\quad\quad\quad} \end{pmatrix}$

115. Let $X = \{1, 2\}$. How many 1–1 mappings of X onto X are there?

Answer: _____

Construct these mappings:

	1 2, 2 1

	1 2, 2 1
	2 1, 1 2

	N	M
N	N	M
M	M	N

	6

	a b c, a c b
	c b a, b a c
	b c a, c a b

	a b c, a c b
	c b a, b a c
	b c a, c a b

$N = \begin{pmatrix} 1 & 2 \\ & \end{pmatrix}$, $M = \begin{pmatrix} 1 & 2 \\ & \end{pmatrix}$

Now compute NN, NM, MN, MM.

$NN = \begin{pmatrix} 1 & 2 \\ & \end{pmatrix}$, $NM = \begin{pmatrix} 1 & 2 \\ & \end{pmatrix}$

$MN = \begin{pmatrix} 1 & 2 \\ & \end{pmatrix}$, $MM = \begin{pmatrix} 1 & 2 \\ & \end{pmatrix}$

Complete the following table:

	N	M
N		
M		

116. Let $X = \{a, b, c\}$. How many 1–1 mappings of X onto X are there?

Answer: _____

List all these mappings below:

$M = \begin{pmatrix} a & b & c \\ & & \end{pmatrix}$, $N = \begin{pmatrix} a & b & c \\ & & \end{pmatrix}$

$L = \begin{pmatrix} a & b & c \\ & & \end{pmatrix}$, $P = \begin{pmatrix} a & b & c \\ & & \end{pmatrix}$

$Q = \begin{pmatrix} a & b & c \\ & & \end{pmatrix}$, $R = \begin{pmatrix} a & b & c \\ & & \end{pmatrix}$

Compute all the products and complete the table below:

$MM = \begin{pmatrix} a & b & c \\ & & \end{pmatrix}$, $MN = \begin{pmatrix} a & b & c \\ & & \end{pmatrix}$

$ML = \begin{pmatrix} a & b & c \\ & & \end{pmatrix}$, $MP = \begin{pmatrix} a & b & c \\ & & \end{pmatrix}$

$MQ = \begin{pmatrix} a & b & c \\ & & \end{pmatrix}$, $MR = \begin{pmatrix} a & b & c \\ & & \end{pmatrix}$

76 Relations and Mappings

$a\,c\,b, a\,b\,c$

$c\,a\,b, b\,c\,a$

$b\,a\,c, c\,b\,a$

$c\,b\,a, b\,c\,a$

$a\,b\,c, c\,a\,b$

$a\,c\,b, b\,a\,c$

$b\,a\,c, c\,a\,b$

$b\,c\,a$

$a\,b\,c, c\,b\,a$

$a\,c\,b, b\,c\,a$

$c\,b\,a, b\,a\,c$

$a\,c\,b, c\,a\,b$

$a\,b\,c, c\,a\,b$

$b\,a\,c, a\,c\,b$

$c\,b\,a, a\,b\,c$

$b\,c\,a$

$NM = \begin{pmatrix} a & b & c \\ & & \end{pmatrix}, \quad NN = \begin{pmatrix} a & b & c \\ & & \end{pmatrix}$

$NL = \begin{pmatrix} a & b & c \\ & & \end{pmatrix}, \quad NP = \begin{pmatrix} a & b & c \\ & & \end{pmatrix}$

$NQ = \begin{pmatrix} a & b & c \\ & & \end{pmatrix}, \quad NR = \begin{pmatrix} a & b & c \\ & & \end{pmatrix}$

$LM = \begin{pmatrix} a & b & c \\ & & \end{pmatrix}, \quad LN = \begin{pmatrix} a & b & c \\ & & \end{pmatrix}$

$LL = \begin{pmatrix} a & b & c \\ & & \end{pmatrix}, \quad LP = \begin{pmatrix} a & b & c \\ & & \end{pmatrix}$

$LQ = \begin{pmatrix} a & b & c \\ & & \end{pmatrix}, \quad LR = \begin{pmatrix} a & b & c \\ & & \end{pmatrix}$

$PM = \begin{pmatrix} a & b & c \\ & & \end{pmatrix}, \quad PN = \begin{pmatrix} a & b & c \\ & & \end{pmatrix}$

$PL = \begin{pmatrix} a & b & c \\ & & \end{pmatrix}$

$PP = \begin{pmatrix} a & b & c \\ & & \end{pmatrix}, \quad PQ = \begin{pmatrix} a & b & c \\ & & \end{pmatrix}$

$PR = \begin{pmatrix} a & b & c \\ & & \end{pmatrix}, \quad QM = \begin{pmatrix} a & b & c \\ & & \end{pmatrix}$

$QN = \begin{pmatrix} a & b & c \\ & & \end{pmatrix}, \quad QL = \begin{pmatrix} a & b & c \\ & & \end{pmatrix}$

$QP = \begin{pmatrix} a & b & c \\ & & \end{pmatrix}, \quad QQ = \begin{pmatrix} a & b & c \\ & & \end{pmatrix}$

$QR = \begin{pmatrix} a & b & c \\ & & \end{pmatrix}, \quad RM = \begin{pmatrix} a & b & c \\ & & \end{pmatrix}$

$RN = \begin{pmatrix} a & b & c \\ & & \end{pmatrix}, \quad RL = \begin{pmatrix} a & b & c \\ & & \end{pmatrix}$

$RP = \begin{pmatrix} a & b & c \\ & & \end{pmatrix}, \quad RQ = \begin{pmatrix} a & b & c \\ & & \end{pmatrix}$

$RR = \begin{pmatrix} a & b & c \\ & & \end{pmatrix}$

Relations and Mappings

Complete the table below:

	M	N	L	P	Q	R
M						
N						
L						
P						
Q						
R						

	M	N	L	P	Q	R
M	M	N	L	P	Q	R
N	N	M	R	Q	P	L
L	L	Q	M	R	N	P
P	P	R	Q	M	L	N
Q	Q	L	P	N	R	M
R	R	P	N	L	M	Q

{(2, 2), (2, 3), (3, 2), (3, 3)}

4

117. Let $X = \{2, 3\}$. Form the set $X \times X$.

$X \times X =$ _____

(a) How many elements are in $X \times X$?

Answer: _____

(b) The counting principle could be used to determine the number of elements in $X \times X$ by reasoning that an ordered pair of $X \times X$ is obtained by completing a task T_1 followed by a task T_2 where T_1 is the task of choosing an element from X to fill the first place of (,) and T_2 is the task of choosing an element from X to fill the second place of (,). This is illustrated below:

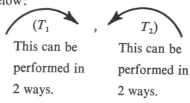

(T_1 , T_2)
This can be performed in 2 ways.   This can be performed in 2 ways.

Therefore, the tasks T_1, T_2 can be performed in

$2 \cdot 2 = 4$ ways and there are 4 elements of $X \times X$.

(c) How many binary operations are there on X? Since each binary operation is a mapping of $X \times X$ to X we may use the following scheme to determine this.

(2, 2) ⟶ ☐   task T_1
(2, 3) ⟶ ☐   task T_2
(3, 2) ⟶ ☐   task T_3
(3, 3) ⟶ ☐   task T_4

The boxes may be filled with any of the elements of X and a binary operation on X will result. Therefore, each task T_1, T_2, T_3, T_4 can be performed in two ways. Then all the boxes can be filled in $2 \cdot 2 \cdot 2 \cdot 2$ ways so that there are 16 binary operations on X.

(d) How many binary operations are there on any set of 2 elements?

Answer: _____

(e) The number of binary operations on $\{2, 3\}$ may also be computed by considering the table and determining the number of ways the boxes can be filled in by using 2 and 3.

	2	3
2	T_1	T_2
3	T_3	T_4

(f) Construct the 16 binary operations on the set $\{2, 3\}$ by filling the tables below:

118. Let $X = \{1, 2, 3\}$.

(a) Determine the number of elements in the set $X \times X$ by using the technique in 117 above. See the scheme suggested below:

80   Relations and Mappings

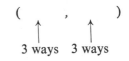

Answer: 9

(b) Determine the number of elements in the set $Y \times X$, where $Y = X \times X$.

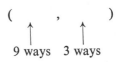

Answer: 27

(c) Determine the number of binary operations on the set X. Using the counting principle and the table below:

	1	2	3
1	T_1	T_2	T_3
2	T_4	T_5	T_6
3	T_7	T_8	T_9

Each T_i, $i = 1, 2, 3, \ldots, 9$ can be performed in 3 ways.

Answer: $3 \times 3 \times 3 \times 3 \times 3 \times 3 \times 3 \times 3 \times 3 = 3^9$

(d) Let B be the binary operation on $\{1, 2, 3\}$ given by the table below:

B	1	2	3
1	2	3	1
2	1	2	3
3	3	1	2

$1B2 =$ 3, $2B1 =$ 1

Is it true that $1B2 = 2B1$?   No

Is B a commutative binary operation?   No

$(1B2)B3 = 3B3 = $ _____

$1B(2B3) = 1B($_____$) = $ _____

Is it true that $(1B2)B3 = 1B(2B3)$? _____

Is B associative? _____

(e) Let B be the binary operation on $\{1, 2, 3\}$ given by the table below:

B	1	2	3
1	1	2	3
2	2	3	1
3	3	1	2

$1B2 = $ _____ ; $2B1 = $ _____

$1B3 = $ _____ ; $3B1 = $ _____

$2B3 = $ _____ ; $3B2 = $ _____

Is B commutative? _____

$(1B1)B2 = $ _____ ; $1B(1B2) = $ _____

$(1B2)B2 = $ _____ ; $1B(2B2) = $ _____

Is B associative? _____

Answers column:

2

3, 1

No

No

2, 2

3, 3

1, 1

Yes

2, 2

3, 3

This cannot be determined on the basis of the above computations.

3 Some Number Theory

This chapter contains a discussion of some basic ideas in *number theory*. This means that we will be discussing facts concerning the natural numbers N:

$$1, 2, 3, 4, 5, \ldots$$

and the integers Z:

$$\ldots -5, -4, -3, -2, -1, 0, 1, 2, 3, 4, 5, \ldots$$

The reader will be familiar with many of the elementary properties of these sets. For example, he will be familiar with the fact that addition, $+$, and multiplication, $\cdot$, are binary operations on Z which are both commutative and associative. These binary operations are related by the *distributive law*:

$$a \cdot (b + c) = a \cdot b + a \cdot c \quad \text{for all } a, b, c \in Z$$

There is a relation, "less than," defined on Z as follows:

$$\{(a, b) \mid a, b \in N \text{ and there is an element } c \in N \text{ such that } a + c = b\}$$

This relation is usually denoted by "$<$" so that the definition takes this form:

$$a < b \text{ if and only if } a + c = b \text{ for some } c \in N$$

For example, $2 < 11$ because $2 + 9 = 11$. We expect the reader to be familiar with the following properties of $<$.

Some Number Theory

For all $u, v \in Z$ one and only one of the following is true:

$$u < v, \quad u = v, \quad v < u \qquad (3\text{-}1)$$

For all $u, v, w \in Z$, $u < v$ if and only if $u + w < v + w$. (3-2)
For all $u, v, w \in Z$ if $w > 0$, then $u < v$ if and only if $u \cdot w < v \cdot w$. (3-3)
For all $u, v, w \in Z$ if $w \neq 0$ and $u \cdot w = v \cdot w$, then $u = v$. (3-4)
For all $u, v, w \in Z$ if $w < 0$, then $u < v$ if and only if $v \cdot w < u \cdot w$. (3-5)

The common practice of writing

$$a < b < c$$

to mean that

$$a < b \quad \text{and} \quad b < c$$

will be followed here. Also we write

$$a \leq b$$

to mean that $a < b$ or $a = b$. The symbol ">" is defined as follows:

$$a > b \text{ if and only if } b < a$$

A detailed development of these topics may be found in the appendix and in B. K. Youse, *The Number System*, Belmont, Calif.: Dickenson Pub. Co., Inc. 1965.

3.1 The Well-Ordering Principle and Induction

A basic characteristic of the set N of natural numbers is the so-called well-ordering property which we assume.

Well-Ordering Property of the Natural Numbers: Let S be a nonempty subset of the natural numbers. Then there is a natural number $t \in S$ such that $t \leq x$ for all $x \in S$.

The number t of this statement is called the *least member* of S. In this language we may say that *every nonvoid set of natural numbers contains a least element.*

Example 3-1: The least natural number in the set N of all natural numbers is 1; 2 is the least element of the set of even natural numbers.

Example 3-2: Show that there are no natural numbers x such that $0 < x < 1$.

Solution: Suppose there is at least one natural number x such that $0 < x < 1$. Then if

$$S = \{x \mid x \text{ is a natural number and } 0 < x < 1\}$$

$S \neq \emptyset$. By the well-ordering property there is a least member, say t, of S. Hence, $0 < t < 1$. If we multiply by t, we have $0 < t^2 < t$ by (3-2) and, hence,

$$0 < t^2 < 1$$

But since the product of natural numbers is a natural number, $t^2 \in S$ with $t^2 < t$. This is a contradiction since t is the smallest member of S and, therefore, proves the statement.

Example 3-3: Prove that if m is any natural number, then $m < 2^m$.

Solution: Suppose that there is some natural number n such that $n \not< 2^n$. Then the set

$$S = \{k \mid k \text{ is a natural number and } k \not< 2^k\}$$

is not empty. So S has a least member, say t. Thus t is the smallest natural number such that

$$t \not< 2^t$$

We note that $t \neq 1$ since $1 < 2^1$. Therefore, $t - 1 < 1$ and $t - 1$ is a natural number smaller than t. Since t is the smallest natural number such that $t \not< 2^t$, we conclude that $t - 1 < 2^{t-1}$. Then, by adding 1, we have $t < 2^{t-1} + 1$. But since $1 < 2^{t-1}$, we may add 2^{t-1} and obtain

$$2^{t-1} + 1 < 2^{t-1} + 2^{t-1} = 1 \cdot 2^{t-1} + 1 \cdot 2^{t-1} = (1 + 1) \cdot 2^{t-1} = 2 \cdot 2^{t-1} = 2^t$$

Hence, $t < 2^t$, contrary to the fact that t was assumed to be the smallest natural number such that $t \not< 2^t$. We conclude that our original supposition that there was at least one natural number n such that $n < 2^n$ is incorrect, and that $m < 2^m$ for all natural numbers m.

In Example 3-3 we have the statement

$$m < 2^m$$

In fact, we have many statements here—one for each natural number m—and we wish to prove that they are all true. It quite often happens that we have a function P which associates with each natural number n a statement $P(n)$ and we wish to show that the set

$$\{P(n) \mid n \text{ is a natural number}\}$$

contains only true statements. In the example above, the statement $P(n)$ for each natural number n was

$$n < 2^n$$

and we were able to prove that these are all true by using the well-ordering property of the natural numbers. Another method of doing this, which is based on the well-ordering property, is called *mathematical induction*.

Some Number Theory

Principle of Mathematical Induction. Let P be a function which associates with each natural number n a statement $P(n)$ and suppose that the following are true:
(a) $P(1)$ is true;
(b) if, for any natural number k, $P(k)$ is true, then $P(k+1)$ is true.

Then $P(n)$ is true for all natural numbers n.

Proof. Suppose that there is at least one natural number s such that $P(s)$ is not true. Then the set

$$S = \{s \mid s \text{ is a natural number and } P(s) \text{ is false}\}$$

is not empty. Let t be the least element of S. Then $1 \notin S$ since by (a) $P(1)$ is true. Thus $t > 1$ and $t - 1$ is a natural number not in S (the smallest element of S is t and $t - 1 < t$). Hence, $P(t - 1)$ is true. By (b), then, $P(t - 1 + 1)$ is true; i.e., $P(t)$ is true. But $t \in S$ so that $P(t)$ is false—a contradiction. This implies that our supposition that there was at least one natural number s such that $P(s)$ is false is incorrect. Therefore, $P(n)$ is true for all natural numbers n.

Let us apply the principle of mathematical induction to the statement

$$n < 2^n$$

The principle states that we can conclude that this statement is true for all natural numbers if we can show that it has properties (a) and (b). It has property (a) since $P(1)$ is the statement

$$1 < 2$$

and this is true. If we know that $P(k)$ is a true statement for a natural number, k, i.e., that

$$k < 2^k$$

then,

$$k + 1 < 2^k + 1 < 2^k + 2^k = 2 \cdot 2^k = 2^{k+1}$$

or $P(k + 1)$ is true. This shows that (b) also is satisfied by P. Hence, we conclude that $P(n)$ is true for all natural numbers.

The procedure we have just used in applying the principle of mathematical induction must be observed generally; i.e., to prove that a statement is true for all natural numbers by mathematical induction we must:
(a) Show that the statement is true for 1.
(b) Show that if the statement is true for a natural number k, then it is true for $k + 1$.

Example 3-4: Let $P(n)$ be the statement

$$n = n + 1$$

Show that P has property (b) but does not have property (a).

Solution: $P(1)$ is the statement

$$1 = 1 + 1$$

which is not true, so that P does not have property (a). To show that P has property (b) we must show that if $P(k)$ is true, then $P(k+1)$ is true. If $P(k)$ is true, then

$$k = k + 1$$

and adding 1 gives

$$k + 1 = (k + 1) + 1$$

Hence, if $k = k + 1$, then $(k + 1) = (k + 1) + 1$. Thus, (b) is satisfied. (The reader has no doubt decided that $n = n + 1$ is false for every natural number n.)

Example 3-5: Show that the statement

$$I(n): \quad n \neq n + 1$$

satisfies both conditions (a) and (b) of the principle of mathematical induction.

Solution: The statement $I(1)$ is

$$1 \neq 1 + 1$$

and this is true, so (a) is satisfied. Now suppose that $k \neq k + 1$. We need to demonstrate that this assumption implies that $k + 1 \neq (k + 1) + 1$. But if we did not have $k + 1 \neq (k + 1) + 1$, we would have $k + 1 = (k + 1) + 1$, and subtracting 1 would give $k = k + 1$. Hence, (b) is satisfied and we may conclude that $n \neq n + 1$ for all natural numbers n.

Example 3-6: Prove that $[n(n + 1)(n + 2)]/3$ is a natural number for all natural numbers n.

Solution: We apply the principle of mathematical induction to the statement $P(n): [n(n + 1)(n + 2)]/3$ is a natural number. For $n = 1$ the statement $P(n)$ is

$$[1(1 + 1)(1 + 2)]/3 \text{ is a natural number}$$

i.e., "2 is a natural number," and this is true, so $P(n)$ satisfies condition (a) of the induction principle. To show that condition (b) is also satisfied, suppose that $P(k)$ is true:

$$[k(k + 1)(k + 2)]/3 \text{ is a natural number}$$

We wish to show that $P(k + 1)$ is true if $P(k)$ is true. The statement $P(k + 1)$ is

$$[(k + 1)(k + 2)(k + 3)]/3 \text{ is a natural number}$$

Using the distributive and commutative axioms we may write

$$(k + 1)(k + 2)(k + 3) = k(k + 1)(k + 2) + 3(k + 1)(k + 2)$$

and therefore,

$$[(k + 1)(k + 2)(k + 3)]/3 = [k(k + 1)(k + 2)]/3 + (k + 1)(k + 2)$$

Since $(k + 1)(k + 2)$ is a natural number, the sum on the right above is a natural number *if* $[k(k + 1)(k + 2)]/3$ is a natural number. In other words, $[(k + 1)(k + 2)(k + 3)]/3$ is a natural number if $[k(k + 1)(k + 2)]/3$ is a natural number. This shows that $P(n)$ satisfies condition (b) and concludes the proof that $[n(n + 1)(n + 2)]/3$ is a natural number for all natural numbers n.

Example 3-7: Prove that for all natural numbers n

$$2 + 4 + 6 + \cdots + 2n = n(n + 1)$$

Solution: We are to prove that the sum of the first n even natural numbers is $n(n + 1)$. The induction principle may be applied to the statement

$$P(n): \quad 2 + 4 + 6 + \cdots + 2n = n(n + 1)$$

$P(1)$ is the statement: $2 \cdot 1 = 1(1 + 1)$. Thus, $P(1)$ is true. The statements $P(k)$ and $P(k + 1)$ are

$$P(k): \quad 2 + 4 + 6 + \cdots + 2k = k(k + 1)$$

and $P(k + 1)$: $2 + 4 + 6 + \cdots + 2k + 2(k + 1) = (k + 1)(k + 2)$. Now suppose that $P(k)$ is true. Then,

$$\begin{aligned} 2 + 4 + 6 + \cdots + 2k + 2(k + 1) &= (2 + 4 + 6 + \cdots + 2k) + 2(k + 1) \\ &= k(k + 1) + 2(k + 1) \\ &= (k + 1)(k + 2) \end{aligned}$$

This shows that if $P(k)$ is true, then so is $P(k + 1)$.

Exercises

(A) In Exercises 1-3 complete the proofs by supplying reasons for the steps.

1. Prove that for $a, b, c \in N$ if $a < b$, then

    $a + c < b + c$.

    Proof.

    1. $a < b$

        Reason: _____ Given

    2. $a + x = b$ for some $x \in N$

        Reason: _____ By Step 1 and definition of $<$

    3. $(a + x) + c = b + c$

        Reason: _____ From Step 2

Some Number Theory   89

4. $a + (x + c) = b + c$

   Reason: _____ From Step 3 and associative property of $+$

5. $a + (c + x) = b + c$

   Reason: _____ From Step 4 and commutative property of $+$

6. $(a + c) + x = b + c$

   Reason: _____ From Step 5 and associative property of $+$

7. $a + c < b + c$

   Reason: _____ From Step 6 and definition of $<$

2. Prove that for $a, b, c \in N$ if $a < b$ and $b < c$, then $a < c$.

   Proof.

   1. $a < b$

      Reason: _____ Given

   2. $a + x = b$ for some $x \in N$

      Reason: _____ Definition of $<$ and Step 1

   3. $b < c$

      Reason: _____ Given

   4. $b + y = c$ for some $y \in N$

      Reason: _____ Step 3 and definition of $<$

   5. $(a + x) + y = c$

      Reason: _____ From Step 2 and Step 4

   6. $a + (x + y) = c$

      Reason: _____ From Step 5 and the associative property of $+$

90 Some Number Theory

From Step 6 and definition of $<$	7. $a < c$ Reason: _____

3. Prove that for $a, b, c, d \in N$ if $a < b$ and $c < d$, then $a + c < b + d$.

   Proof.

Given	1. $a < b$ Reason: _____
From Exercise 1 above and Step 1	2. $a + c < b + c$ Reason: _____
Given	3. $c < d$ Reason: _____
From Exercise 1 above and Step 3	4. $b + c < b + d$ Reason: _____
From Exercise 2 above and Steps 2 and 4	5. $a + c < b + d$ Reason: _____

(B) In Exercises 4–8 a general statement $P(n)$ is given, where n is a natural number. The blanks are to be filled in with the appropriate statement corresponding to the given values of n.

4. *Sample.* If $P(n)$ is the statement
   $0 \leq n^2 - 2n\sqrt{n} + n$ then
   $P(5)$ is $\underline{0 \leq 25 - 10\sqrt{5} + 5}$
   $P(4)$ is $\underline{0 \leq 16 - 8 \cdot \sqrt{4} + 4}$
   $P(3)$ is $\underline{0 \leq 12 - 6\sqrt{3}}$
   $P(2)$ is $\underline{0 \leq 6 - 4\sqrt{2}}$
   $P(1)$ is $\underline{0 \leq 0}$

Some Number Theory   91

$P(k+1)$ is $\underline{0 \leq (k+1)^2 - 2(k+1)\sqrt{k+1} + (k+1)}$

5. If $P(n)$ is the statement

   $[n(n+1)]/2$ is a natural number, then $P(12)$ is

   _____

78 is a natural number

45 is a natural number $P(9)$ is _____

15 is a natural number $P(5)$ is _____

3 is a natural number $P(2)$ is _____

1 is a natural number $P(1)$ is _____

$[k(k+1)]/2$ is a natural number $P(k)$ is _____

$[(k+1)(k+2)]/2$ is a natural number $P(k+1)$ is _____

$[k(2k+1)]$ is a natural number $P(2k)$ is _____

6. If $P(n)$ is the statement

   $1 + 2 + 3 + \cdots + n = [n(n+1)]/2$ then $P(7)$ is

   _____

$1 + 2 + 3 + 4 + 5 + 6 + 7 = 28$

$1 + 2 + 3 + 4 = 10$ $P(4)$ is _____

$1 + 2 + 3 = 6$ $P(3)$ is _____

$1 + 2 = 3$ $P(2)$ is _____

$1 = 1$ $P(1)$ is _____

$1 + 2 + 3 + \cdots + k = [k(k+1)]/2$ $P(k)$ is _____

$1 + 2 + 3 + \cdots + k + (k+1)$ $P(k+1)$ is _____

$= [(k+1)(k+2)]/2$

_____

7. If $P(n)$ is the statement

   $1 + 4 + 9 + \cdots + n^2 = [n(n+1)(2n+1)]/6$

$1 + 4 + 9 + 16 + 25 + 36 + 49$ then $P(8)$ is _____

$+ 64 = [8(9)(17)]/6$ _____

$1 + 4 + 9 + 16 + 25 = [5(6)(11)]/6$ $P(5)$ is _____

92 Some Number Theory

$1 + 4 + 9 = 14$

$1 + 4 = 5$

$1 = 1$

$1 + 4 + 9 + \cdots + a^2 =$
$[a(a + 1)(2a + 1)]/6$

$1 + 4 + 9 + \cdots + a^2 + (a + 1)^2$
$= [(a + 1)(a + 2)(2a + 3)]/6$

$P(3)$ is _____

$P(2)$ is _____

$P(1)$ is _____

$P(a)$ is _____

_____

$P(a + 1)$ is _____

_____

8. If $P(n)$ is the statement
$$2 + 2(2^2) + 3(2^3) + \cdots + n(2^n) = 2 + (n - 1) \cdot 2^{n+1}$$
then $P(5)$ is _____

$2 + 2(2^2) + 3(2^3) + 4(2^4) + 5(2^5)$
$= 2 + 4 \cdot 2^6$

$2 + 2(2^2) + 3(2^3) + 4(2^4)$
$= 2 + 3 \cdot 2^5$

$2 + 2(2^2) + 3(2^3) = 2 + 2 \cdot 2^4$

$2 + 2(2^2) = 2 + 1 \cdot 2^3$

$2 = 2 + 0 \cdot 2^2$

$2 + 2(2^2) + 3(2^3) + 4(2^4) + \cdots +$
$t(2^t) = 2 + (t - 1)2^{t+1}$

$2 + 2(2^2) + 3(2^3) + 4(2^4) + \cdots +$
$t(2^t) + (t + 1)(2^{t+1}) = 2 + t \cdot 2^{t+2}$

$P(4)$ is _____

_____

$P(3)$ is _____

$P(2)$ is _____

$P(1)$ is _____

$P(t)$ is _____

_____

$P(t + 1)$ is _____

_____

(B) In Exercises 9–15, proofs by mathematical induction are given. You are to fill in the blanks so that the proofs become complete.

9. *Sample*. Prove that $\dfrac{n(n + 3)}{2}$ is a natural number for all natural numbers n.

Solution. We let $P(n)$ be the statement

Some Number Theory   93

$\frac{n(n+3)}{2}$ is a natural number.

and apply the Principle of Mathematical Induction. We first check the statement $P(1)$. Since $P(1)$ is the statement

$\frac{1(1+3)}{2}$ is a natural number.

$P(1)$ is true. Now we wish to show that $P(k+1)$ is true if $P(k)$ is true. Now $P(k)$ and $P(k+1)$ are the following statements;

$P(k)$: $\frac{k(k+3)}{2}$ is a natural number

$P(k+1)$: $\frac{(k+1)(k+4)}{2}$ is a natural number.

We may use the distributive axiom to write $[(k+1)(k+4)]/2$ as follows:

$$\frac{(k+1)(k+4)}{2} = \frac{k(k+4)+(k+4)}{2}$$

$$= \frac{k(k+3+1)+(k+4)}{2}$$

$$= \frac{k(k+3)+k+(k+4)}{2}$$

$$= \frac{k(k+3)}{2} + \frac{2k+4}{2}$$

$$= \frac{k(k+3)}{2} + \frac{2(k+2)}{2}$$

$$= \frac{k(k+3)}{2} + \underline{\hspace{2cm}} \quad (k+2)$$

From this we may conclude that

$$\frac{(k+1)(k+4)}{2} = \frac{k(k+3)}{2} + \underline{\hspace{2cm}} \quad (k+2)$$

Since $k+2$ is a natural number, this last equation

94 Some Number Theory

$\dfrac{n^5}{5} + \dfrac{n^3}{3} + \dfrac{7n}{15}$ is a natural number

$\dfrac{1}{5} + \dfrac{1}{3} + \dfrac{7}{15}$ is a natural number

1, true

$P(k+1)$ is true

$\dfrac{k^5}{5} + \dfrac{k^3}{3} + \dfrac{7k}{15}$ is a natural number

$\dfrac{(k+1)^5}{5} + \dfrac{(k+1)^3}{3} + \dfrac{7(k+1)}{15}$ is a natural number.

$k^4 + 2k^3 + 2k^2 + k + \dfrac{1}{5}$

shows that if $P(k)$ is true, i.e., *if $[k(k+3)]/2$ is a natural number*, then $[(k+1)(k+4)]/2$ is a natural number or $P(k+1)$ is true. Therefore, by the Principle of Mathematical Induction we have shown that the statement $P(n)$ is true for all natural numbers.

10. Prove that for all natural numbers n
$$\dfrac{n^5}{5} + \dfrac{n^3}{3} + \dfrac{7n}{15}$$
is a natural number.

Solution. Let $P(n)$ be the statement

_____

Then $P(1)$ is

_____

Since $\dfrac{1}{5} + \dfrac{1}{3} + \dfrac{7}{15} = $ _____, $P(1)$ is _____.

Now we wish to show that if $P(k)$ is true, then _____. The statement $P(k)$ is _____

and $P(k+1)$ is _____

_____

Let us simplify the arithmetic in $P(k+1)$;

(a) $\dfrac{(k+1)^5}{5} = \dfrac{k^5 + 5k^4 + 10k^3 + 10k^2 + 5k + 1}{5}$

$= \dfrac{k^5}{5} + $ _____

Some Number Theory   95

$$= \left(\frac{k^5}{5} + \frac{1}{5}\right) + (\underline{})$$

(b) $\dfrac{(k+1)^3}{3} = \dfrac{k^3 + 3k^2 + 3k + 1}{3}$

$$= \left(\frac{k^3}{3} + \frac{1}{3}\right) + (\underline{})$$

(c) $\dfrac{7(k+1)}{15} = \underline{\phantom{\dfrac{7k}{15} + \dfrac{7}{15}}}$

Then from (a), (b), (c) we conclude that

$\dfrac{(k+1)^5}{5} + \dfrac{(k+1)^3}{3} + \dfrac{7(k+1)}{15} = \dfrac{k^5}{5} + \dfrac{1}{5} +$

$\underline{\phantom{\dfrac{k^3}{3} + \dfrac{1}{3} + \dfrac{7k}{15} + \dfrac{7}{15}}}$

$+ k^4 + \underline{}$

$+ k^2 + k = \dfrac{k^5}{5} + \dfrac{k^3}{3} + \dfrac{7k}{15} + (\underline{})$

We know that $k^4 + 2k^3 + 3k^2 + 2k + 1$ is a

\underline{} and if $P(k)$ is

true, $\dfrac{k^5}{5} + \dfrac{k^3}{3} + \dfrac{7k}{15}$, \underline{}

is a natural number. Thus if $P(k)$ is true,

$\dfrac{(k+1)^5}{5} + \dfrac{(k+1)^3}{3} + \dfrac{7(k+1)}{15}$ is the sum of two

\underline{} and is

therefore a natural number. Hence, if $P(k)$ is true, $P(k+1)$ is true. This completes the proof by induction.

11. Prove that for all natural numbers n
$2 + 2\cdot 3 + 2\cdot 3^2 + \cdots + 2\cdot 3^{n-1} = 3^n - 1$.

Solution: Let $P(n)$ be the statement
$2 + 2\cdot 3 + 2\cdot 3^2 + \cdots + 2\cdot 3^{n-1} = 3^n - 1$.

96   Some Number Theory

$2 = 3^1 - 1$

true

$2 + 2 \cdot 3 + 2 \cdot 3^2 + \cdots + 2 \cdot 3^{k-1}$
$= 3^k - 1$

$2 + 2 \cdot 3 + 2 \cdot 3^2 + \cdots + 2 \cdot 3^{k-1} +$
$2 \cdot 3^k = 3^{k+1} - 1$

$2 + 2 \cdot 3 + 2 \cdot 3^2 + \cdots + 2 \cdot 3^{k-1} +$
$2 \cdot 3^k$

$2 \cdot 3^k$

$3^k - 1$

$2 + 1$

$3^{k+1} - 1$

5 is a factor of 5

5 is a factor of $8^k - 3^k$

5 is a factor of $8^{k+1} - 3^{k+1}$

$8^k, 3^k$

Then $P(1)$ is _____

and this is a _____ statement.

We now wish to show that if $P(k)$ is true, then $P(k+1)$ is true also. $P(k)$ is _____

_____

and $P(k+1)$ is _____

_____

Now if $P(k)$ is true, we have _____

_____

$= (2 + 2 \cdot 3 + 2 \cdot 3^2 + \cdots + 2 \cdot 3^{k-1}) +$ _____

$= ($ _____ $) + 2 \cdot 3^k$ (Use the truth of $P(k)$ here)

$= 2 \cdot 3^k + 3^k - 1 = 3^k($ _____ $) - 1$

$= 3^k \cdot 3 - 1 =$ _____

This shows that if $P(k)$ is true, then $P(k+1)$ is true and completes the proof.

12. Show that for all natural numbers n, 5 is a factor of $8^n - 3^n$.

    Solution. Let $P(n)$ be the statement 5 is a factor of $8^n - 3^n$

    $P(1)$ is the statement _____

    and this is true. Now we try to show that if $P(k)$ is true, then so is $P(k+1)$. $P(k)$ is

    _____

    and $P(k+1)$ is _____

    We may express $8^{k+1} - 3^{k+1}$ as follows:

    $8^{k+1} - 3^{k+1} = 8 \cdot$ _____ $- 3 \cdot$ _____

    $= (5 + 3) \cdot 8^k - 3 \cdot 3^k = 5 \cdot 8^k +$

Some Number Theory   97

$3 \cdot 8^k, 3 \cdot 3^k$

_____ − _____ $= 5 \cdot 8^k + 3 \cdot (8^k - 3^k)$

Now if $P(k)$ is true we may write $8^k - 3^k = 5a$ where a is some natural number. Then we have

$8^{k+1} - 3^{k+1} = 5 \cdot 8^k + 3 \cdot (8^k - 3^k) = 5 \cdot 8^k + 3(5a)$

$8^k + 3a$

$= 5($ _____ $)$

$8^{k+1} - 3^{k+1}$

Thus if $P(k)$ is true, 5 is a factor of _____ i.e., $P(k+1)$ is true. Therefore, the proof is complete.

13. Prove that if n and b are natural numbers, then there are integers q, r such that $n = bq + r$ and $0 \leq r < b$.

Solution. Let $P(n)$ be the statement

_____

_____

There are integers q, r such that $n = bq + r$ and $0 \leq r < b$.

Then $P(1)$ is the statement _____

There are integers q, r such that $1 = bq + r$ and $0 \leq r < b$.

If $b = 1$, we may write $1 = 1 \cdot 1 + 0$ and $0 \leq 0 < 1$ so that in this case q, r may be taken to be 1, 0 respectively. If $1 < b$, we may write $1 = b \cdot 0 + 1$ and $0 \leq 1 < b$ so that q, r may be taken to be 0, 1. In either case we see that $P(1)$ is _____.

true

Now we show that _____,

if $P(k)$ is true

_____.

then $P(k+1)$ is true

$P(k)$ is the statement _____

There are integers q, r such that $k = bq + r$ and $0 \leq r < b$.

But if $k = bq + r$, adding 1 gives

$k + 1 = bq + (r + 1)$

_____.

Some Number Theory

P(k + 1) is true
bq
b, q + 1
k + 1 = b(q + 1) + 0

If $r + 1 < b$ we have that there are integers q, $r + 1$ such that $k + 1 = bq + (r + 1)$ and $0 \leq r + 1 < b$. But this says that _____ in this case. If $r + 1 \not< b$, then $r + 1 = b$ so that

$$k + 1 = bq + r + 1 = \underline{} + \underline{} = b \cdot (\underline{}) + 0$$

This shows that in this case there are integers $q + 1$, 0 such that _____ and $0 \leq 0 < b$ and again $P(k + 1)$ is true if $P(k)$ is true. This completes the induction proof.

(D) Prove each of the following, Exercises 14–20.

14. If $a, b, c \in N$ and $a < b$, then $a \cdot c < b \cdot c$. (Hint: Use Exercise 1 above as a model.)

15. For $a, b, c, d \in N$

   if $a < b$ and $c < d$, then $a \cdot c < b \cdot d$

16. If $n \in N$, then $\dfrac{n(n + 1)}{2}$ is a natural number.

17. If $n \in N$, then $\dfrac{n(n + 1)(n + 2)}{6}$ is a natural number.

18. If $n \in N$, then

$$1 + 2 + 3 + \cdots + n = \dfrac{n(n + 1)}{2}$$

19. If $n \in N$, then

$$1 + 3 + 5 + \cdots + (2n - 1) = n^2$$

20. If $n \in N$, then
$$1 + 5 + 9 + \cdots + (4n - 3) = n(2n - 1)$$

21. Prove that the number of lines connecting n points in a plane, where no 3 of the points are on a line, is $[n(n - 1)]/2$ where n is any natural number such that $n \geq 2$.

22. Prove that for every natural n, $n^3 + (n + 1)^3 + (n + 2)^3$ has a factor of 9.

23. Prove the following generalization of the induction principle: Let $P(n)$ be a statement that is associated with every natural number n such that $n \geq a$ where a is a natural number. Suppose the following is true:
    (a) $P(a)$ is true.
    (b) If $P(k)$ is true for any natural number $k \geq a$, $P(k + 1)$ is true.
    Then $P(n)$ is true for all natural numbers $n \geq a$.

24. Use the well-ordering property of the natural numbers to prove the following:
    Let Z be a subset of the natural numbers N which satisfies the following two conditions:
    (a) $1 \in Z$.
    (b) If $k \in Z$, then $(k + 1) \in Z$.
    Then $Z = N$.

25. Prove that if $a, b, c \in N$ and $ac = bc$, then $a = b$.

26. Prove that if $n, b \in N$, then the following is not possible: $0 < nb < b$.

3.2 Division and the Division Algorithm

The definition of $<$ for integers is as follows: for $a, b \in Z$

$a < b$ if and only if there is an element $c \in N$ such that $a + c = b$

Observe that $a < b$ is defined in terms of $+$. If we change the definition by replacing $+$ by $\cdot$ and replacing N by Z we obtain the definition of "divides" as follows:

For integers a, b we say that a *divides* b if and only if there is an element $c \in Z$ such that $a \cdot c = b$.

If *a* divides *b* we write $a\,|\,b$ and say that *b is a multiple of a*. For example, $6\,|\,24$ because $6 \cdot 4 = 24$; 24 is a multiple of 6.

Theorem 6. Let $a, b, c \in Z$. Then
 (a) $a\,|\,a$;
 (b) If $a\,|\,b$ and $b\,|\,c$, then $a\,|\,c$;
 (c) If $a\,|\,b$ and $a\,|\,c$, then $a\,|\,(b+c)$;
 (d) If $a\,|\,b$, then $a\,|\,bc$.

Proof. (a) For any $a \in Z$ we have $a \cdot 1 = a$. So by definition of "divides," $a\,|\,a$.
 (b) If $a\,|\,b$ and $b\,|\,c$, there are elements $x, y \in Z$ such that $ax = b$ and $by = c$, by the definition of "divides." Replace b by ax in the second equation and we obtain

$$(ax)y = c \text{ or } a(xy) = c$$

Then, since xy is an integer, $a\,|\,c$ by definition.
 (c) If $a\,|\,b$ and $a\,|\,c$, there are numbers $x, y \in Z$ such that $ax = b$ and $ay = c$. Then

$$b + c = ax + ay = a(x + y)$$

and by the definition of "divides," $a\,|\,(b+c)$.
 (d) If $a\,|\,b$, then there is an element x such that $b = ax$. Then $bc = axc = a(xc)$ and $a\,|\,bc$ by definition.

Suppose that $n, b \in N$. If $b \nmid n$ (*b* does not divide *n*) then *n* is not a multiple of *b*, i.e., *n* is not a member of the set

$$\{0, b, 2b, 3b, 4b, \ldots, qb, (q+1)b, \ldots\} = \{qb\,|\,q \in N \text{ or } q = 0\}$$

It seems reasonable, however, that *n* should lie *between* two multiples of *b*:

$$qb < n < (q+1)b \text{ for some } q = 0, 1, 2, 3, \ldots$$

or

$$qb < n < qb + b$$

Then from this last pair of inequalities, subtracting qb gives

$$0 < n - qb < b$$

Now let $n - qb = r$. We may then write

$$n = qb + r \text{ with } 0 < r < b \tag{3-4}$$

It was shown in Exercise 13 above that for $n, b \in N$ there are integers q, r such that

$$n = qb + r \text{ with } 0 \leq r < b \tag{3-5}$$

Statement (3-5) is exactly the same as (3-4) except that in (3-5) it is possible for r to be 0. This exception allows *n* to be a multiple of *b*.

Some Number Theory

Theorem 7. (*Division Algorithm.*) If $n, b \in N$, there are integers q, r such that
$$n = bq + r \text{ and } 0 \leq r < b$$
The integers q and r are unique.

Proof. The existence of the integers q, r has already been established in Exercise 13 above. It only remains to prove the uniqueness of q and r. Therefore, suppose that besides q, r there are integers q_1, r_1 such that $n = bq_1 + r_1$ and $0 \leq r_1 < b$. Then we have

$$n = bq + r \text{ and } 0 \leq r < b$$

and

$$n = bq_1 + r_1 \text{ and } 0 \leq r_1 < b$$

Then

$$bq + r = bq_1 + r_1$$

If $r < r_1$, then $0 < r_1 - r < b$ and from
$$bq + r = bq_1 + r_1$$
$$0 < r_1 - r = bq - bq_1 = b(q - q_1) < b$$

This says that a multiple of b lies between 0 and b, which is impossible (see Exercise 26). Therefore, it is not possible for $r < r_1$. In the same way we may show that we cannot have $r_1 < r$. By (3-1) the only remaining possibility is $r_1 = r$. Then from $bq + r = bq_1 + r_1$ we conclude that

$$bq = bq_1$$

and (by Exercise 25) $q = q_1$. Thus the integers q, r are unique.

We describe the Division Algorithm by saying that we may divide a natural number n by the natural number b to obtain a quotient q and a remainder r—which is less than b.

Example 3-8: In the Division Algorithm, take $b = 2$. Then for any $n \in N$
$$n = 2q + r \text{ and } 0 \leq r < 2$$
Since $0 \leq r < 2$, the only possibilities for r are 0, 1. Hence,
$$n = 2q + 0 = 2q$$
or
$$n = 2q + 1$$
This, of course, is just another way of saying that every natural number is even ($r = 0$) or odd ($r = 1$).

Example 3-9: By Example 3-8 each natural number n may be written as follows:
$$n = 2q \text{ or } n = 2q + 1$$

Then
$$n^2 = 4q^2 \text{ or } n^2 = (2q+1)^2 = 4q^2 + 4q + 1 = 4(q^2 + q) + 1$$

The Division Algorithm states that the possible remainders when dividing by 4 are 0, 1, 2, 3. The above equations show that a square (n^2) divided by 4 leaves only the remainders 0, 1.

Exercises

(A) Give a reason for the truth of each statement in Exercises 27–35.

27. $4 < 8$ because _____

$4 + 4 = 8$

28. $4 \mid 8$ because _____

$4 \cdot 2 = 8$

29. For $x \in N$, $x < 2x$ because _____

$x + x = 2x$

30. For $x \in Z$, $x \mid 2x$ because _____

$x \cdot 2 = 2x$

31. For $t \in N$, $2 \mid 2^t$ because _____

$2 \cdot 2^{t-1} = 2^t$

32. For $a \in Z$, $a \mid a^2$ because _____

$a \cdot a = a^2$

33. $8 \mid 0$ because _____

$8 \cdot 0 = 0$

Some Number Theory 103

$a \cdot 0 = 0$

$p^2 \mid p^2$ and $p^2 \mid p^3$ or $p^2 \cdot (1 + p)$
$= p^2 + p^3$

34. For $a \in Z$, $a \mid 0$ because _____

35. For $p \in N$, $p^2 \mid (p^2 + p^3)$ because _____

(B) In Exercises 36–42 you are given natural numbers a, b. In each case find integers q, r with $0 \leq r < b$ such that $a = bq + r$.

36. Sample. $a = 29$, $b = 3$
    Answer: $29 = 3 \cdot 9 + 2$

37. $a = 28$, $b = 5$
    Answer: _____

$28 = 5 \cdot 5 + 3$

38. $a = 5$, $b = 28$
    Answer: _____

$5 = 28 \cdot 0 + 5$

39. $a = 105$, $b = 36$
    Answer: _____

$105 = 36 \cdot 2 + 33$

40. $a = 1023$, $b = 93$
    Answer: _____

$1023 = 93 \cdot 11 + 0$

41. $a = 4162$, $b = 29$
    Answer: _____

$4162 = 29 \cdot 143 + 15$

42. a is a square, $b = 4$

Some Number Theory

$a = 4q + 0$ or $a = 4q + 1$

$n = 2q + 1$, for some natural number q

$n^2 = 4q^2 + 4q + 1 = 4(q^2 + q) + 1$

$q = 2k$, for some $k \in N$

$q = 2k + 1$, for some $k \in N$

$n^2 = 4(4k^2 + 2k) + 1$
$= 8(2k^2 + k) + 1$

$n^2 = 4(q^2 + q) + 1$
$= 4((4k^2 + 4k + 1)$
$+ (2k + 1) + 1$
$= 4(4k^2 + 6k + 2) + 1$
$= 8(2k^2 + 3k + 1) + 1$

If n is an odd natural number, then

Answer: _____

(See Example 3-9, p. 108.)

(C) Supply the requested information in Exercises 43-48.

43. (a) If n is an odd natural number, how may n be expressed?

Answer: _____
_____

(b) If $n = 2q + 1$, compute n^2.

Answer: _____

(c) The number q of (a), (b) is either even or odd. If it is even, how may it be expressed?

Answer: _____

If it is odd, how may it be expressed?

Answer: _____

(d) If $q = 2k$, compute n^2 from (b) in terms of k.

Answer: _____
_____

(e) If $q = 2k + 1$, compute n^2 from (b) in terms of k.

Answer: _____
_____
_____
_____

(f) From (a)-(e) above, what conclusion can be drawn about the square of an odd natural number?

Answer: _____

its square, n^2, may be expressed as $8t + 1$ for some $t \in N$.

$n + 1, n + 2$

$n = 3q$ or $n = 3q + 1$ or $n = 3q + 2$ for some $q \in N$

$3q, 3q + 1, 3q + 2$

$3q + 1, 3q + 2, 3q + 3 = 3(q + 1)$

$3q + 2, 3q + 3 = 3(q + 1), 3q + 4$
$= (3q + 3) + 1 = 3(q + 1) + 1$

They may be expressed as $3t, 3t + 1,$ $3t + 2$ or as $3t + 1, 3t + 2,$ $3(t + 1)$ or as $3t + 2, 3(t + 1),$ $3(t + 1) + 1$ for some $t \in N$.

44. (a) If $n \in N$, what are the next two consecutive natural numbers after n?
Answer: _____

(b) Use the Division Algorithm and divide n by 3. How can n then be expressed?
Answer: _____

(c) If $n = 3q$, how can $n, n + 1, n + 2$ be expressed in terms of q?
Answer: _____

(d) If $n = 3q + 1$, how can $n, n + 1, n + 2$ be expressed in terms of q?
Answer: _____

(e) If $n = 3q + 2$, how can $n, n + 1, n + 2$ be expressed?
Answer: _____

(f) From (a)-(e), what conclusions can be drawn about any three consecutive natural numbers?
Answer: _____

(g) Since $3t + 1$ and $3t + 2$ are consecutive, what can be concluded about at least one of them?

At least one of them is even.

$(3t+1)(3t+2)$ is even or
$2|(3t+1)(3t+2)$

Since $3|(3t)$ and $2|(3t+1)(3t+2)$ we conclude that $6|(3t)(3t+1)(3t+2)$ or that 6 is a factor of the product of any 3 consecutive natural numbers.

Answer: _____

(h) From (g), what can be concluded about $(3t+1)(3t+2)$?

Answer: _____
_____

(i) From (h), what can be concluded about $(3t)(3t+1)(3t+2)$?

Answer: _____
_____
_____
_____
_____

This is another way of proving that $\dfrac{(3t)(3t+1)(3t+2)}{6}$ is a natural number.

See Exercise 17.

45. Prove that the product of four consecutive natural numbers is a multiple of 24. Use Exercise 44 as a model.

46. Can you use Exercise 44 to prove that for any $n \in N$, $6|(n^3 - n)$? Explain.

47. Prove that for any $n \in N$, $2|(n^2 + n)$.

$x = 2t + 1, y = 2s + 1$ for some $s, t \in N$

48. (a) If $x \in N$, $y \in N$ and both x, y are odd, how can x and y be expressed?

Answer: _____
_____

(b) If x, y are both odd, use the expression of x, y obtained in (a) to compute $x^2 + y^2$.

Some Number Theory   107

$x^2 + y^2 = (2t + 1)^2 + (2s + 1)^2$
$= (4t^2 + 4t + 1) + (4s^2 + 4s + 1)$
$= 4(t^2 + t + s^2 + s) + 2$
$= 2[2(t^2 + t + s^2 + s) + 1]$

If $x, y \in N$ and both x, y are odd, then $2 | (x^2 + y^2)$ or the sum of the squares of two odd natural numbers has a factor of 2.

Answer: _____
_____
_____
_____

(c) From (a) and (b), what conclusion can be drawn?

Answer: _____
_____
_____
_____

(d) In (b) above, it was shown that if $x, y \in N$ and both are odd, then $x^2 + y^2 = 4q + 2$ for some $q \in N$. Using the Division Algorithm assertion that quotients and remainders are unique we may conclude that remainders of 0, 1, 3 cannot be obtained by dividing $x^2 + y^2$ by 4. In particular, we conclude that if $x, y \in N$ and both are odd, then $x^2 + y^2$ is not a multiple of 4.

49. Prove the following for $a, b, c, x, y \in Z$.
   (a) If $a | b$, then $a | bc$;
   (b) If $a | b$ and $a | c$, then $a | (bx + cy)$;
   (c) If $a | b$ and $a | c$, then $a | (bx - cy)$.

3.3 Greatest Common Divisors and Least Common Multiples

We now are ready to consider the basic concepts of GCD's and LCM's (*greatest common divisors* and *least common multiples*, respectively).

DEFINITION.   Let $a, b \in Z$. The natural number c is the GCD of a, b if and only if the following is true:

(1) $c\,|\,a$ and $c\,|\,b$ (c is a common divisor of a, b);
(2) If $x\,|\,a$ and $x\,|\,b$ for $x \in Z$, then $x\,|\,c$.

The notation GCD (a, b) will be used to denote the GCD of a, b.

DEFINITION. Let $a, b \in Z$. The natural number c is the LCM of a, b, if and only if the following is true:
(1) $a\,|\,c$ and $b\,|\,c$ (c is a common multiple) of a, b;
(2) If $a\,|\,x$ and $b\,|\,x$ for $x \in Z$, then $c\,|\,x$.

The notation LCM(a, b) will be used in the obvious way.

Example 3-9: For the integers 16 and 24 we see that the following is the set of common divisors:

$$\{\pm 1, \pm 2, \pm 4, \pm 8\}$$

The largest of these numbers is 8. Observe that each of the common divisors of 16, 24 divides 8 and hence that 8 is the GCD of 16, 24. The LCM of 16 and 24 is 48 since $48 = 16 \cdot 3$ and $48 = 24 \cdot 2$ and 48 is the "smallest" multiple of both 16 and 24.

In Examples 3-10 and 3-11 below we obtain GCD's by the method known as the Euclidean Algorithm. It should be clear from these examples that this method of producing GCD's relies heavily on the following property of "divides" (see Exercise 40):

$$\text{If } a\,|\,b \text{ and } a\,|\,c, \text{ then } a\,|\,(bx \pm cy)$$

Example 3-10: Suppose we wish to compute GCD(24, 238). In the left-hand column below we used the Division Algorithm three times to obtain the 3 equations. On the right-hand side we have described the process which occured on the left in terms of arbitrary natural numbers a, b.

(1) $238 = 9 \times 24 + 22, 0 \leq 22 < 24$ | $a = q \times b + r, 0 \leq r < b$
(2) $24 = 1 \times 22 + 2, 0 \leq 2 < 22$ | $b = q_1 \times r + r_1, 0 \leq r_1 < r$
(3) $22 = 1 \times 2 + 0, 0 \leq 0 < 2$ | $r = q_2 \times r_1 + r_2, 0 \leq r_2 < r_1$
$0 < 2 < 22 < 24$ | $r_2 = 0 < r_1 < r < b$

From these equations we conclude that GCD(24, 238) = 2 and GCD$(a, b) = r_1$. This may be seen as follows:

from (3), $2\,|\,22$ | $r_1\,|\,r$
from (2), $2\,|\,24$ | $r_1\,|\,b$ (since $r_1\,|\,r_1$ and $r_1\,|\,r$)
(since $2\,|\,2$ and $2\,|\,22$) |
from (1), $2\,|\,238$ | $r_1\,|\,a$ (since $r_1\,|\,r$ and $r_1\,|\,b$)
(since $2\,|\,22$ and $2\,|\,24$) |
Hence, $2\,|\,24$ and $2\,|\,238$. | Hence $r_1\,|\,a$ and $r_1\,|\,b$.
And so by definition, 2 is a common divisor of 24 and 238. If $c\,|\,238$ and | And so by definition, r_1 is a common divisor of a and b. If $c\,|\,a$ and $c\,|\,b$, then

$c \mid 24$, $c \mid 22$ [from (1)]; if $c \mid 24$ and $c \mid 22$, $c \mid 2$ [from (2)]. Thus, if c is a common divisor of 238 and 24, c is a divisor of 2; i.e., 2 is the GCD of 238 and 24 by definition.

$c \mid r$ [from (1)]; if $c \mid b$ and $c \mid r$, then $c \mid r_1$ [from (2)]. Thus, if c is a common divisor of a and b, c is a divisor of r_1, i.e., r_1 is the GCD of a and b by definition.

Observe that the GCD was obtained by the above calculations when the last non-zero remainder was obtained.

Example 3-11: We use the method described above in Example 3-10 to compute GCD(288, 456).

$456 = 1 \times 288 + 168$
$288 = 1 \times 168 + 120$
$168 = 1 \times 120 + 48$
$120 = 2 \times 48 + 24$
$48 = 2 \times 24 + 0$
$0 < 24 < 48 < 120 < 168$
GCD(456, 288) = 24

$a = q \times b + r, 0 \leq r < b$
$b = q_1 \times r + r_1, 0 \leq r_1 < r$
$r = q_2 \times r_1 + r_2, 0 \leq r_2 < r_1$
$r_1 = q_3 \times r_2 + r_3, 0 \leq r_3 < r_2$
$r_2 = q_4 \times r_3 + r_4, 0 \leq r_4 < r_3$
$0 = r_4 < r_3 < r_2 < r_1 < r$
GCD$(a, b) = r_3$

It seems intuitively clear that every pair of integers a, b has a GCD. This needs to be established, however, and a proof will be given in Theorem 9. First, we need another result.

Theorem 8. Let S be a nonvoid set of integers which has the following property:

$$\text{if } a, b \in S, \text{ then } a - b \in S$$

Then either $S = \{0\}$ or there is a positive integer p such that $S = \{np \mid n \in Z\}$.

Proof. Suppose that $S \neq \{0\}$. Then there is some $x \in S$ such that $x \neq 0$. Applying the hypothesis that $a - b \in S$ if $a, b \in S$ we have $x - x = 0 \in S$ and $0 - x = -x \in S$. One of the two integers x and $-x$ is positive since $x \neq 0$. Then the set $P = \{y \mid y \in S \text{ and } y \text{ is positive}\}$ is not empty. By the well-ordering principle P has a smallest element, say p. We now wish to show that S contains all np where $n \in N$. This may be done by applying induction to the statement $Q(n)$ below:

$$Q(n): \quad n \cdot p \in S$$

$Q(1)$ is the statement $1 \cdot p \in S$. But this is true since $p \in P \subseteq S$. Now suppose that $Q(k)$ is true, i.e., suppose $k \cdot p \in S$. Since $p \in S$, $0 \in S$, $0 - p \in S$, i.e., $-p \in S$. Hence, since $k \cdot p \in S$ and $-p \in S$, $k \cdot p - (-p) \in S$. But $k \cdot p - (-p) = kp + p = (k + 1)p$. Therefore $(k + 1) \cdot p \in S$ if $k \cdot p \in S$. This shows that $Q(k + 1)$ is true if $Q(k)$ is true. Thus we conclude that S contains all $n \cdot p$ for $n \in N$. S also contains $0 \cdot p = 0$ from above and also

$$0 - n \cdot p = (-n) \cdot p \in S \text{ for any } n \in N$$

Therefore, $\{n \cdot p \mid n \in Z\} \subseteq S$. If $x \in S$, we may use the Division Algorithm and

write $x = m \cdot p + r$, $0 \leq r < p$. Apply the hypothesis to the numbers x and $m \cdot p$ (both are in S) and obtain

$$r = x - m \cdot p \in S$$

Since $r < p$, if $r \neq 0$ the choice of p as the smallest positive element of S is contradicted. Hence, $r = 0$ and $x = m \cdot p$. Then $S \subseteq \{n \cdot p \mid n \in Z\}$ and this completes the proof that $S = \{n \cdot p \mid n \in Z\}$.

Theorem 9. Let $a, b \in Z$. Then a and b have a GCD if not both of a, b is zero.

Proof. Let $S = \{ax + by \mid x \in Z, y \in Z\}$. Let us choose elements of S; $ax_1 + by_1$ and $ax_2 + by_2$ for some $x_1, y_1, x_2, y_2 \in Z$. Then $(ax_1 + by_1) - (ax_2 + by_2) = a(x_1 - x_2) + b(y_1 - y_2) \in S$. This shows that S satisfies the hypothesis of Theorem 8. Also $S \neq \emptyset$ since $a = a \cdot 1 + b \cdot 0 \in S$ and $b = a \cdot 0 + b \cdot 1 \in S$. $S \neq \{0\}$ because not both a, b is zero. We may then apply Theorem 8 and conclude that S consists of all $n \cdot p$ for $n \in Z$ and positive $p \in S$. Since a and b are both in $S(a = a \cdot 1 + b \cdot 0$ and $b = a \cdot 0 + b \cdot 1)$, we conclude that $a = m \cdot p$, $b = q \cdot p$ for some $m, q \in Z$. Hence, $p \mid a$ and $p \mid b$. By definition of S, there are integers s, t such that $p = a \cdot s + b \cdot t$. Then if $c \mid a$ and $c \mid b$, $c \mid as$ and $c \mid bt$. Therefore, $c \mid (as + bt)$; i.e., $c \mid p$. We conclude that $p = \text{GCD}(a, b)$ by definition of GCD.

The proof of Theorem 9 is referred to in mathematics as an "existence proof," because the proof shows that there *is* a GCD for each pair of integers a, b but there is no indication of how one constructs it. An *algorithm* is a systematic procedure for doing something. The procedure described above in Examples 3-10 and 3-11 is an algorithm for constructing the GCD of two integers.

Observe that in Theorem 9, $p \in S$. By definition of S this implies that there are integers x, y such that

$$p = ax + by$$

An expression like $ax + by$ is called a *linear combination* of a, b. We may say, therefore, that GCD(a, b) *may be expressed as a linear combination of a, b.*

Example 3-12: In Example 3-10 the following computations were used to compute GCD(238, 24):

(1) $238 = 9 \cdot 24 + 22$
(2) $24 = 1 \cdot 22 + 2$
(3) $22 = 11 \cdot 2 + 0$

These equations may be used to exrpess GCD(238, 24) (which is 2) as a linear combination of 238, 24. From (2)

$$2 = 24 - 1 \cdot 22$$

and from (1)

$$22 = 238 - 9 \cdot 24$$

Now substitute this expression of 22 into the equation $2 = 24 - 1 \cdot 22$:

$$2 = 24 - 1 \cdot 22$$
$$= 24 - 1 \cdot (238 - 9 \cdot 24)$$
$$= 24 - 1 \cdot 238 + 9 \cdot 24$$
$$= (-1) \cdot 238 + 10 \cdot 24$$

Therefore, $2 = (238) \cdot (-1) + 24 \cdot 10$ and 2, the GCD of 238 and 24, has been expressed as a linear combination of 238 and 24; i.e., in the form $2 = 238x + 24y$ where $x = -1$ and $y = 10$.

Example 3-13: In Example 3-11 we found that GCD(456, 288) = 24 by using the following computations:

$$456 = 1 \cdot 288 + 168$$
$$288 = 1 \cdot 168 + 120$$
$$168 = 1 \cdot 120 + 48$$
$$120 = 2 \cdot 48 + 24$$

From these we may obtain the following (starting from the last):
$$24 = 120 - 2 \cdot 48$$
(a) $\quad 48 = 168 - 1 \cdot 120$
(b) $\quad 120 = 288 - 1 \cdot 168$
(c) $\quad 168 = 456 - 1 \cdot 288$

Now, starting with $24 = 120 - 2 \cdot 48$ we make successive substitutions:

$$24 = 120 - 2 \cdot 48$$
$$= 120 - 2(168 - 1 \cdot 120) \quad \text{from (a)}$$
$$= 120 - 2 \cdot 168 + 2 \cdot 120$$
$$= 3 \cdot 120 - 2 \cdot 168$$
$$= 3 \cdot (288 - 1 \cdot 168) - 2 \cdot 168 \quad \text{from (b)}$$
$$= 3 \cdot 288 - 3 \cdot 168 - 2 \cdot 168$$
$$= 3 \cdot 288 - 5 \cdot 168$$
$$= 3 \cdot 288 - 5 \cdot (456 - 1 \cdot 288) \quad \text{from (c)}$$
$$= 3 \cdot 288 - 5 \cdot 456 + 5 \cdot 288$$
$$= 8 \cdot 288 - 5 \cdot 456$$

Therefore, $24 = (456)(-5) + 288 \cdot 8$ and GCD(456, 288) (which is 24) has been expressed as a linear combination of 456 and 288.

Exercises

In Problems 50–67 compute the GCD of the given integers and then express the GCD as a linear combination of the two integers.

50. (a) Compute GCD(60, 50).

$60 = 50 \cdot$ _____ 1, 10 _____ $+$ _____

$50 = 10 \cdot$ _____ 5 _____

Therefore, GCD(60, 50) = _____ 10 _____

(b) Now use part (a) to find integers x, y such that $10 = 60x + 50y$.

$10 = 60 \cdot$ _____ 1, (−1) _____ $+ 50 \cdot$ _____

Therefore, $x =$ _____ 1, −1 _____ and $y =$ _____

51. (a) Compute GCD(250, 100).

$250 = 100 \cdot$ _____ 2, 50 _____ $+$ _____

$100 = 50 \cdot$ _____ 2 _____

Therefore, GCD(250, 100) = _____ 50 _____

(b) Use part (a) to find integers x, y such that $50 = 250x + 100y$.

$50 = 250 \cdot$ _____ 1, −2 _____ $+ 100 \cdot$ _____

Hence, $x =$ _____ 1, −2 _____ and $y =$ _____

52. (a) Compute GCD(1024, 24).

$1024 = 24 \cdot$ _____ 42, 16 _____ $+$ _____ (a.1)

$24 = 16 \cdot$ _____ 1, 8 _____ $+$ _____ (a.2)

$16 = 8 \cdot$ _____ 2 _____

Thus, GCD(1024, 24) = _____ 8 _____

(b) Use part (a) to find integers such that $8 = 1024x + 24y$. From (a.2) above

$8 = 24 \cdot$ _____ 1, −1 _____ $+ 16 \cdot$ _____ (a.3)

and from (a.1)

$16 = 1024 - 24 \cdot 42$ (a.4)

Now substitute for 16 in (a.3):

Some Number Theory   113

1024 − 24·42

1024, 24

43

−1, 43

$8 = 24 \cdot 1 + (\underline{}) \cdot (-1)$

$= 24 \cdot 1 + \underline{} \cdot (-1) + \underline{} \cdot 42$

$= 1024 \cdot (-1) + 24 \cdot \underline{}$

Therefore, $8 = 1024 \cdot (-1) + 24 \cdot 43$ so that

$x = \underline{}$ and $y = \underline{}$

53. (a) Compute GCD(2856, 784).

3, 504

1, 280

1, 224

1, 56

4

56

$2856 = 784 \cdot \underline{} + \underline{}$ (b.1)

$784 = 504 \cdot \underline{} + \underline{}$ (b.2)

$504 = 280 \cdot \underline{} + \underline{}$ (b.3)

$280 = 224 \cdot \underline{} + \underline{}$ (b.4)

$224 = 56 \cdot \underline{}$ (b.5)

Therefore, GCD(2856, 784) = \underline{}

(b) Use part (a) to find integers x, y such that

224

$2 = 2856x + 284y$. From (b.4) $56 = 280 − \underline{}$.

Substitute from (b.3) for 224 into $56 = 280 − 224$;

504 − 280

2

$56 = 280 − (\underline{})$

$= 280 − 504 + 280 = 280 \cdot \underline{} − 504$

Substitute for 280 from (b.2):

784 − 504

2, 2

$56 = 280 \cdot 2 − 504 = (\underline{}) \cdot 2 − 504$

$= 784 \cdot \underline{} − 504 \cdot \underline{} − 504$

$= 784 \cdot 2 − 504 \cdot 3$

Substitute for 504 from (b.1):

2856 − 784·3

3, 9

11, −3

$56 = 784 \cdot 2 − 504 \cdot 3 = 784 \cdot 2 − (\underline{}) \cdot 3$

$= 784 \cdot 2 − 2856 \cdot \underline{} + 784 \cdot \underline{}$

$= 784 \cdot \underline{} + 2856 \cdot \underline{}$

Therefore, $x = −3$ and $y = 11$.

54. Compute GCD(1246, 124) and find integers x, y

2, 21, −211

1, −9, 137

1, −8, 37

1, −332, 1275

$\{24n \mid n \in Z\}$
24

such that GCD(1246, 124) = 1246x + 124y.
Answer: GCD = _____, x = _____, y = _____

55. Compute GCD(487, 32) and find integers x, y such that GCD(487, 32) = 487x + 32y.
Answer: GCD = _____, x = _____, y = _____

56. Compute GCD(615, 133) and find integers such that GCD(615, 133) = 615x + 133y.
Answer: GCD = _____, x = _____, y = _____

57. Compute GCD(8107, 2111) and find integers x, y such that GCD(8107, 2111) = 8107x + 2111y.
Answer: GCD = _____, x = _____, y = _____

58. The set
$\{6n \mid n \in Z\} = \{0, \pm 6, \pm 12, \pm 18, \pm 24, \ldots\}$
is the set of multiples of 6 and
$\{8n \mid n \in Z\} = \{0, \pm 8, \pm 16, \pm 24, \pm 32, \ldots\}$
is the set of multiples of 8. What set is the intersection of these two?
$\{6n \mid n \in Z\} \cap \{8n \mid n \in Z\} =$ _____
Therefore, LCM(6, 8) = _____

59. If an integer $t (\neq 0)$ is a common multiple of both 6 and 8, what may be concluded? It may be concluded that there are integers x, y such that $t = 6x = 8y$. From this it follows that $2^3 \mid t$ and $3 \mid t$. Hence $24 \mid t$ [see Exercise 49(a)]. Thus, every common multiple of 6 and 8 is also a multiple of 24.

24

72

90

550

4620

$GCD(a, b) = ax + by$ for some $x, y \in Z$

$t \cdot (ax + by) = tax + tby$

$GCD(a, b) = ax + by$ for some $x, y \in Z$

Therefore, LCM(6, 8) = _____

60. Compute LCM(24, 36).
    Answer: _____

61. Compute LCM(18, 45).
    Answer: _____

62. Compute LCM(55, 50).
    Answer: _____

63. Compute LCM(420, 330).
    Answer: _____

64. Prove that for $a, b, t \in Z$ and $t > 0$, $GCD(ta, tb) = t \cdot GCD(a, b)$.

    Proof. First, express $GCD(a, b)$ as a linear combination of a, b: _____

    Then $t \cdot GCD(a, b) = $ _____
    Since $(ax + by) | a$ and $(ax + by) | b$, $t \cdot (ax + by) | ta$ and $t \cdot (ax + by) | t \cdot b$. Therefore, $t \cdot GCD(a, b)$ is a common divisor of ta and tb. And if $c | ta$ and $c | tb$, $c | (tax + tby) = t \cdot GCD(a, b)$; i.e., $GCD(ta, tb) = t \cdot GCD(a, b)$.

65. Prove that if $a | bc$ and $GCD(a, b) = 1$, then $a | c$.

    Proof. Write $GCD(a, b)$ as a linear combination of a, b: _____

It is given that $GCD(a, b) = 1$ so that $1 = ax + by$. Now multiply this by c: $c = cax + bcy$. Since $a\,|\,cax$ and $a\,|\,bcy$ (given), $a\,|\,(cax + bcy)$ or $a\,|\,c$.

66. Prove that if $a\,|\,b$ and $a\,|\,c$, then $GCD(b/a, c/a) = (1/a) \cdot GCD(b, c)$.

67. Prove that $GCD(b/a, c/a) = 1$ if $a = GCD(b, c)$.

3.4 The Fundamental Theorem of Arithmetic

Every natural number a may be written as $a \cdot 1$. Hence, every natural number except 1 has at least two natural number factors. *A natural number p is prime if and only if it has exactly two natural number factors.* Some examples of primes are 2, 3, 5, 7, 11, 13, 19, 23. The primes are extremely important in the theory of numbers because they are in a sense the "building blocks" of natural numbers (with respect to multiplication).

Theorem 10. (*Fundamental Theorem of Arithmetic*). Every natural number $a \neq 1$ is either prime or may be written as a product of primes.

Proof. If a is prime, we are done. If a is not prime, we may write $a = b \cdot c$ where both $b, c \neq 1$ and both $b, c \neq a$. Then we must have

$$1 < b < a \text{ and } 1 < c < a$$

If both b, c are prime, we are done. If either b or c is not prime, we repeat this process and continue until we reach only prime factors. The process must terminate in primes only since the factors are "getting smaller" (and by the well-ordering principle every set of natural numbers has a least member).

It seems natural to inquire about the number of primes. This was supposedly answered by Euclid, and the following proof is attributed to him.

Theorem 11. There are infinitely many primes.

Proof. Suppose to the contrary that there are only a limited number of primes. Then there is a largest prime, say P. We may then list *all* the primes:

$$2, 3, 5, 7, 11, \ldots, P$$

Consider the number
$$N = (2 \cdot 3 \cdot 5 \cdot 7 \cdot 11 \cdot \ldots \cdot P) + 1$$
By Theorem 10, N is prime or may be written as a product of primes. In either case there is a prime p such that $p \mid N$. But since $2, 3, 5, 7, 11, \ldots, P$ are all the primes, p must be one of these. However, the definition of N shows that if N is divided by any one of the numbers $2, 3, 5, 7, 11, \ldots, P$ the remainder is 1 (not zero). Thus $p \nmid N$. Therefore, we have a contradiction and the theorem is proved.

Theorem 12. If p is prime and $a \in Z$, then either $\text{GCD}(p, a) = p$ or $\text{GCD}(p, a) = 1$. Moreover, $\text{GCD}(p, a) = p$ only if $p \mid a$.

Proof. Since p is prime, it has only two factors, namely 1 and p. Hence, there are only two possibilities for $\text{GCD}(p, a)$. If $\text{GCD}(p, a) = p$, then by definition of GCD, $p \mid a$.

Theorem 13. If p is a prime and $p \mid ab$, then $p \mid a$ or $p \mid b$.

Proof. Suppose that $p \nmid a$. Then by Theorem 12 $\text{GCD}(p, a) = 1$. Now express $\text{GCD}(p, a)$ as a linear combination of p, a:
$$1 = px + ay \text{ for some } x, y \in Z$$
Multiplying by b:
$$b = pxb + aby$$
By hypothesis $p \mid ab$ and since $p \mid pxb$, $p \mid b$.

The result just obtained (Theorem 13) is a special case of Exercise 65. In that exercise the hypothesis included that $\text{GCD}(a, b) = 1$. Notice that this information was obtained immediately in the proof of Theorem 13 because of the fact that $p \nmid a$.

DEFINITION. Integers a, b are said to be *relatively prime* if and only if $\text{GCD}(a, b) = 1$.

Example 3-13: The hypothesis of Theorem 13 that p is prime is essential, as the following shows:
$$6 \text{ is not prime, } 6 \mid 12 \text{ but } 6 \nmid 3 \text{ and } 6 \nmid 4$$

Exercises

Mark Exercises 68–77 T or F.

68. F _____ . $3 \mid (3^2 + 1)$

69. F _____ . $3 \mid (3^{101} + 1)$

F · 70. _____. $6 \mid (2 \cdot 4 \cdot 6 \cdot 8 \cdot 10 + 1)$

F · 71. _____. $k \mid (2 \cdot 3 \cdot 4 \cdot 5 \ldots k) + 1$

T · 72. _____. If $7 \mid 100 \cdot a$, then $7 \mid a$ (see Exercise 65)

T · 73. _____. If $33 \mid 100 \cdot a$, then $33 \mid a$ (see Exercise 65)

F · 74. _____. If $33 \mid 3a$, then $33 \mid a$

T · 75. _____. If $97 \mid 36x$, then $97 \mid x$ (see Theorem 13)

T · 76. _____. If $95 \mid 32y$, then $95 \mid y$

F · 77. _____. If $95 \mid 19w$, then $95 \mid w$

78. Write each of the following numbers as products of primes.

$2 \cdot 2 \cdot 3 \cdot 3$ (a) $36 = $ _____
$3 \cdot 3 \cdot 5$ (b) $45 = $ _____
$2 \cdot 2 \cdot 5 \cdot 5$ (c) $100 = $ _____
$3 \cdot 5 \cdot 7$ (d) $105 = $ _____
$3 \cdot 3 \cdot 3 \cdot 3 \cdot 11$ (e) $891 = $ _____
$3 \cdot 5 \cdot 7 \cdot 13 \cdot 17$ (f) $23205 = $ _____

79. Find the first 10 smallest positive integers x such that $\text{GCD}(26, x) = 1$.

1, 3, 5, 7, 9, 11, 15, 17, 19, 21

Some Number Theory 119

1, 2, 4, 7, 8, 11, 13, 14	**80.** Find all positive integers y such that $y < 15$ and $\text{GCD}(15, y) = 1$.
1, 2, 3, 4, 5, 6, 7, 8, 9, 10, 11, 12, 13, 14, 15, 16	**81.** Find all positive integers z such that $z < 17$ and $\text{GCD}(17, z) = 1$.
1, 5, 7, 11, 13, 17, 19, 23	**82.** Find all positive integers w such that $w < 24$ and $\text{GCD}(24, w) = 1$.
2, 5 3, 7 No prime factor of one number is a prime factor of the other.	**83.** The numbers 10, 21 are relatively prime. List the prime factors of 10. Answer: _____ List the prime factors of 21. Answer: _____ How do the prime factors of 10 and 21 compare? Answer: _____
No	**84.** If two integers a, b have a prime factor in common, is it possible to have $\text{GCD}(a, b) = 1$?
	85. Suppose that p is a prime factor of an integer a, i.e., $a = p \cdot x$ for some $x \in Z$. What can be said about the number of times which p occurs as a

Some Number Theory

> *p* occurs an even number of times.
>
> *a* is a perfect square.
>
> By 84 above, x, y have no common prime factors and by 85 the prime factors of xy occur an even number of times. Therefore, the prime factors of x and also y must occur an even number of times as factors of x and y respectively. Therefore, by 86 both x and y are serfect spuares.

factor of $a^2 = a \cdot a$? _____

86. Suppose that a is a positive integer and that each prime factor of a occurs an even number of times. What can be said about a? _____

87. Suppose that x, y are positive integers, xy is a perfect square and that $\mathrm{GCD}(x, y) = 1$. What may be concluded about x, y? _____
_____
_____
_____
_____
_____
_____
_____

88. Find two natural numbers x, y such that neither x nor y is a square but $x \cdot y$ is a square.

89. Let p be a prime and let a_i $(i = 1, 2, 3, \ldots, n)$ be integers. Prove that if $p \mid a_1 \cdot a_2 \cdot a_3 \ldots a_n$, then $p \mid a_i$ for some i.

90. Suppose that for some integer t, the integer $3t + 1$ is a prime. Show that there is an integer n such that $3t + 1 = 6n + 1$.

91. Suppose that a, b are relatively prime integers. Prove that if $a \mid t$ and $b \mid t$, then $ab \mid t$.

92. Show by an example that the hypothesis that a, b be relatively prime in 91 above is necessary.

3.5 Congruences

DEFINITION. Let $a, b, m \in Z$ and $m > 1$. We say that a is *congruent* to b *modulo* m provided $m | (a - b)$.

Standard notations for this concept are:

$$a \equiv b \text{ modulo } m$$

$$a \equiv b \ (m)$$

The second of these will be used here.

Example 3-14:
(a) $3 \equiv 1 \ (2)$ because $2 | (3 - 1)$
(b) $15 \equiv 3 \ (6)$ because $6 | (15 - 3)$
(c) $-18 \equiv 10 \ (7)$ because $7 | (-18 - 10)$

The similarity in appearance between $a \equiv b \ (m)$ and $a = b$ has no doubt occured to the reader. This similarity is more than a superficial one as the following results will show. To clarify the discussion we remind the reader that $a = b$ means that *a and b are names for the same thing*. Using this definition one easily proves the following:
(i) $a = a$
(ii) If $a = b$, then $b = a$.
(iii) If $a = b$ and $b = c$, then $a = c$.

This shows, of course, that "=" is an equivalence relation. Now, using the definition of "congruence, modulo m," we can prove that it too is an equivalence relation.

Theorem 14. "Congruence, modulo m" is an equivalence relation on the set Z, i.e.,
(i′) $a \equiv a \ (m)$ for all $a \in Z$
(ii′) If $a \equiv b \ (m)$, then $b \equiv a \ (m)$ for all $a, b \in Z$
(iii′) If $a \equiv b \ (m)$ and $b \equiv c \ (m)$, then $a \equiv c \ (m)$ for all $a, b, c \in Z$.

Proof. (i′) Since $a - a = 0$ and $m | 0$, we have $a \equiv a \ (m)$ by definition.
(ii′) If $a \equiv b \ (m)$, then $m | (a - b)$ by definition. Hence, there is $k \in Z$ such that $m \cdot k = a - b$. Multiplying by -1 gives

$$m \cdot (-k) = b - a$$

or $m | (b - a)$. Then by definition of "congruence, modulo m" $b \equiv a \ (m)$.
(iii′) If $a \equiv b \ (m)$ and $b \equiv c \ (m)$, then $m | (a - b)$ and $m | (b - c)$. Then there are integers x, y such that

$$m \cdot x = a - b \text{ and } m \cdot y = b - c$$

Adding these gives

$$m \cdot x + m \cdot y = (a - b) + (b - c) = a - c$$

or $m \cdot (x + y) = a - c$. Therefore, $m \mid (a - c)$ and $a \equiv c \ (m)$.

Before presenting further evidence of the likeness of these two concepts we prove the following result which allows us to speak of "congruence, modulo m" in an equivalent manner.

Theorem 15. For $a, b \in Z$, $a \equiv b \ (m)$ if and only if a, b have the same remainder when divided by m.

Proof. (a) Suppose that a, b have the same remainder, r, when divided by m. Then there are integers x, y such that

$$a = mx + r$$
$$b = my + r \quad \text{and} \quad 0 \leq r < m$$

Then $a - mx = b - my$ since each of these is equal to r. Therefore, $a - b = mx - my = m(x - y)$, $m \mid (a - b)$ and, hence, $a \equiv b(m)$.

(b) Now suppose that $a \equiv b(m)$ and divide both a and b by m. There are integers x, y, r, s such that

$$a = mx + r, \quad 0 \leq r < m$$
$$b = my + r, \quad 0 \leq s < m$$

Then $a - b = (mx + r) - (my + s)$
$ = m(x - y) + (r - s)$.

Since $a \equiv b(m)$, we have $m \mid (a - b)$ and $m \mid m(x - y)$. Therefore, $m \mid (r - s)$ and $m \cdot k = r - s$ for some $k \in Z$. If $k \geq 1$, then $m \cdot k \geq m$ and if $k \leq -1$, then $mk \leq -m$. But since $0 \leq r < m$ and $0 \leq s < m$, we must have

$$-m < r - s < m$$

Hence, the integer k is not such that $k \geq 1$ or $k \leq -1$. The only remaining possibility is $k = 0$. Then from $m \cdot k = r - s$, $0 = r - s$ and $r = s$; i.e., a, b have the same remainder when divided by m.

For integers a, b, c we know that the following is true:

$$\text{if } a = b, \text{ then } a + c = b + c \text{ and } a \cdot c = b \cdot c$$

The following is a similar result for congruence.

Theorem 16. Let a, b, c be integers. Then if $a \equiv b(m)$

$$a + c \equiv b + c(m) \text{ and } a \cdot c \equiv b \cdot c(m)$$

Proof. If $a \equiv b(m)$, then $m \mid (a - b)$. But $a - b = (a + c) - (b + c)$ so that $m \mid (a + c) - (b + c)$ and $a + c \equiv b + c(m)$. Also if $m \mid (a - b)$, $m \mid (a - b)c$ or $m \mid (ac - bc)$ and again $a \cdot c \equiv b \cdot c(m)$.

Some Number Theory   123

Corollary.   If $a \equiv b(m)$ and $c \equiv d(m)$, then
$$a + c \equiv b + d(m) \text{ and } a \cdot c \equiv b \cdot d(m)$$

Proof.   Applying Theorem 16 to $a \equiv b(m)$ we have $a + c \equiv b + c(m)$; applying it to $c \equiv d(m)$ we have $b + c \equiv b + d(m)$. Then from
$$a + c \equiv b + c(m) \text{ and } b + c \equiv b + d(m)$$
we obtain $a + c \equiv b + d(m)$ by Theorem 14 (iii′). The other part is proved in a similar way.

Example 3-15:   If n is an even integer, $n \equiv 0(2)$ since n and 0 leave the same remainder when divided by 2.

Example 3-16:   If n is any integer, then
$$n = 3q$$
$$n = 3q + 1$$
or
$$n = 3q + 2$$
for some integer q. Thus, either
$$3\,|\,(n - 0),\; 3\,|\,(n - 1),\; \text{or}\; 3\,|\,(n - 2)$$
i.e.,
$$n \equiv 0(3),\; n \equiv 1(3) \text{ or } n \equiv 2(3)$$

Example 3-17:   The result in Example 3-16 may be generalized. Suppose n is any integer and m is an integer > 1. Then the possible remainders when n is divided by m are $0, 1, 2, 3, \ldots, m - 1$, i.e.,
$$n = m \cdot q + r,\; 0 \leq r < m$$
Therefore, $n - r = mq$ and $n \equiv r(m)$. This shows that *every integer is congruent modulo m to some nonnegative integer less than m.*

Example 3-18:   Using the corollary to Theorem 16 we can obtain from $16 \equiv 10(6)$ and $122 \equiv 62(6)$ that $16 \cdot 122 \equiv 10 \cdot 62(6)$ or $1952 \equiv 620(6)$. Also $122 + 16 \equiv 62 + 10(6)$ or $138 \equiv 72(6)$.

Exercises

$8\,|\,16$

$2\,|\,(-4)$

93. $18 \equiv 2\;(8)$ because _____

94. $-1 \equiv 3\;(2)$ because _____

124 Some Number Theory

$2 \mid 2a$

$9 \mid 99$

$4 \mid 16$

$4 \mid 4a^2$

$15 \mid (x-4)$

$82 \mid (y^2 - 13)$

3

8, 13, 18

4

10, 16, 22

4

−9, −22, −35

10

95. $2a \equiv 0$ (2) because _____

96. $100 \equiv 1$ (9) because _____

97. $35 \equiv 19$ (4) because _____

98. $4a^2 + 1 \equiv 1$ (4) because _____

99. If $x \equiv 4$ (15), then _____

100. If $y^2 \equiv 13$ (82), then _____

101. (a) Find the smallest positive solution for the congruence $2x \equiv 1$ (5). _____
     (b) Find the next three larger solutions to $2x \equiv 1$ (5). _____

102. (a) Find the smallest positive solution for the congruence $5x \equiv 2$ (6). _____
     (b) Find the next three larger solutions to $5x \equiv 2$ (6). _____

103. (a) Find the smallest positive solution for the congruence $10x \equiv 1$ (13). _____
     (b) Find the next 3 smaller solutions for $10x \equiv 1$ (13). _____

104. (a) Find the smallest positive solution for the congruence $10x \equiv 9$ (13). _____

Some Number Theory 125

−3, −16, −29

(b) Find the next 3 smaller solutions for $10x \equiv 9\ (13)$. _____

105. Supply the missing steps in the following proof.

Theorem 17. If a, m are relatively prime integers, then there is some integer x such that $a \cdot x \equiv 1\ (m)$.

Proof. Since a, m are relatively prime, their GCD is _____1_____. Then 1 may be expressed as _____a linear combination of a, m_____; $1 = ax + my$ for some integers x, y. Then from this, $ax - 1 = m(-y)$ and hence $m\ |\ $_____$(ax-1)$_____ or $ax \equiv $ _____1_____ (m).

106. Supply the missing steps in the following proof.

Theorem 18. If a, m are relatively prime and b is any integer, then there is an integer k such that $a \cdot k \equiv b\ (m)$.

Proof. Since a, m are relatively prime we may use Theorem 17 and conclude that _____there is an integer x such that $ax \equiv 1\ (m)$_____. We may now use Theorem 16 and multiply $a \cdot x \equiv 1\ (m)$ by b to obtain _____$a \cdot (x \cdot b) \equiv b\ (m)$_____ Therefore, $x \cdot b$ is the integer k which we wished to find.

107. Supply the missing steps in the following proof.

Theorem 19. If c and m are relatively prime integers and if $a \cdot c \equiv b \cdot c\ (m)$, then $a \equiv b\ (m)$.

Proof. Since c and m are relatively prime we may

as a linear combination of c, m

$a = acx + amy$, $b = bcx + bmy$

$(ac - bc)x + (a - b)my$

express 1 _____ ;
$1 = cx + my$ for some integers x, y. Now multiply this by a and b respectively to obtain
_____ and _____
Then $a - b = $ _____
By hypothesis $ac \equiv bc\ (m)$ or $m\,|\,(ac - bc)$. Therefore, in the above equation $m\,|\,(ac - bc)x$ and $m\,|\,(a - b)my$ so $m\,|\,(a - b)$, i.e., $a \equiv b\ (m)$.

108. Prove by induction that $10^n \equiv 1\ (9)$ for all natural numbers n.

109. Prove that if $a \cdot c \equiv b\ (m)$, then $a \cdot (c + mx) \equiv b\ (m)$ for any integer x.

110. The integer 3851 may be written as follows:
$$3851 = 1 + 5 \cdot 10 + 8 \cdot 100 + 3 \cdot 1000$$
$$= 1 + 5 \cdot 10 + 8 \cdot 10^2 + 3 \cdot 10^3$$

In fact, any positive integer m may be written in this fashion:
$$m = a_0 + a_1 \cdot 10 + a_2 \cdot 10^2 + \cdots + a_n 10^n$$

Here the a_i ($i = 1, 2, \ldots, n$) are the digits of the integer and $0 \leq a_i \leq 9$. It is now possible to show that $m \equiv a_0 + a_1 \cdot a_2 + \cdots + a_n\ (9)$, i.e., $a_0 + a_1 \cdot 10 + a_2 \cdot 10^2 + \cdots + a_n \cdot 10^n \equiv a_0 + a_1 + a_2 + \cdots + a_n\ (9)$. By Exercise 108, $10^n \equiv 1\ (9)$. By Theorem 16 this last congruence may be multiplied by a_n:
$$a_n \cdot 10^n \equiv a_n\ (9)$$

Using Exercise 108 and Theorem 16 as often as is necessary we obtain the following sequence of congruences:

$$a_n \cdot 10^n \equiv a_n\ (9)$$
$$a_{n-1} \cdot 10^{n-1} \equiv a_{n-1}\ (9)$$
$$\vdots$$
$$a_3 \cdot 10^3 \equiv a_3\ (9)$$
$$a_2 \cdot 10^2 \equiv a_2\ (9)$$
$$a_1 \cdot 10 \equiv a_1\ (9)$$
$$a_0 \equiv a_0\ (9)$$

Now use the corollary to Theorem 16 to add all these:
$$a_0 + a_1 \cdot 10 + a_2 \cdot 10^2 + \cdots + a_n \cdot 10^n \equiv a_0 + a_1 + a_2 + \cdots + a_n \ (9)$$
Using the result that two integers are congruent modulo 9 if and only if they leave the same remainder when divided by 9 (this is Theorem 15) we see that an integer m has the same remainder when divided by 9 as the sum of its digits, $a_0 + a_1 + a_2 + \cdots + a_n$. In the example above, $m = 3851$. The sum of the digits here is $3 + 8 + 5 + 1 = 17$. Since the remainder, when 17 is divided by 9, is 8 we can say that 3851 has a remainder of 8 when divided by 9. This result is known as "casting out 9's." Use this result to determine the remainders when each of the following numbers is divided by 9:

(a) 101
(b) 856
(c) 10231
(d) 5280
(e) 2763
(f) 827145
(g) 10^{100}
(h) 981726354

4 Groups

In Chapter 2, binary operations on sets were introduced. Commutative and associative binary operations have also been mentioned. Among the most important binary operations are those called *groups*.

4.1 Definitions and Examples

DEFINITION. Let G be a set and let B be a binary operation on G; i.e., $B: G \times G \to G$. The pair (G, B) is called a *group* if and only if each of the following is true:

(1) B is associative on G,
$$xB(yBz) = (xBy)Bz \text{ for all } x, y, z \in G$$

(2) There is an element of G called the identity (with respect to B) and denoted by e such that
$$xBe = x \text{ and } eBx = x \text{ for all } x \in G$$

(3) For each element $g \in G$ there is a corresponding element h called an *inverse* of g (with respect to B) such that
$$gBh = e \text{ and } hBg = e$$

Example 4-1: Let $G = \{0, 1\}$ and let B be the binary operation given by the table below:

B	0	1
0	0	1
1	1	0

This binary operation was shown to be associative in Chapter 2. Also, 0 is the identity, e, since $0B0 = 0$, $0B1 = 1$, $1B0 = 1$. The inverse of 0 is 0 and the inverse of 1 is 1:

$$0B0 = 0 \text{ and } 1B1 = 0$$

Example 4-2: Let $X = \{1, 2\}$ and let G be the set of all 1–1 mappings of X onto X. The reader will recall the there are 2 of these mappings (see Chapter 2, Problem 104, p. 65). These are $I = \begin{pmatrix} 1 & 2 \\ 1 & 2 \end{pmatrix}$ and $A = \begin{pmatrix} 1 & 2 \\ 2 & 1 \end{pmatrix}$. The set G is then $\{I, A\}$. The product, $\cdot$, of mappings is a binary composition on $\{I, A\}$ which may be given by the following table:

$\cdot$	I	A
I	I	A
A	A	I

Then the pair $(G, \cdot)$ is a group with identity I and each element is its own inverse. Compare this with Example 4-1.

Example 4-3: Let G be the set of positive rational numbers, all positive numbers of the form a/b where a, b are positive integers. Ordinary multiplication, $\times$, on this set is a binary operation (another way of saying this is that the product of two positive rational numbers is a positive rational number). Then the pair $(G, \times)$ is a group: multiplication is associative, $1 = 1/1$ is the identity element and b/a is the inverse of a/b since $a/b \times b/a = 1/1 = 1$.

Example 4-4: Let $G = \{-1, 1\}$ and let B be ordinary multiplication, $\times$, on G. B may be given by the following table:

$\times$	1	-1
1	1	-1
-1	-1	1

The pair $(G, \times)$ is a group. Multiplication is associative, 1 is the identity element and each element is its own inverse. Compare this with Examples 4-1 and 4-2.

Example 4-5: Let $G = \{5x \mid x \text{ is an integer}\}$ and let B be ordinary addition. Addition on G is a binary operation since if $5x, 5y \in G$, then
$$5x + 5y = 5(x + y)$$
and $5(x + y) \in G$. Since $0 = 5 \cdot 0 \in G$, 0 is an identity:
$$5x + 0 = 5x \text{ and } 0 + 5x = 5x$$
For $5x \in G$, the element $5(-x) \in G$ is its inverse:
$$5x + 5(-x) = 5(x + (-x)) = 5 \cdot 0 = 0$$
and also $5(-x) + 5x = 0$. Since addition of integers is associative, $(G, +)$ is a group.

Consider the equilateral triangle in Fig. 4-1.

Figure 4-1

This triangle may be rotated about the center point 0 in either the clockwise or counterclockwise direction. We wish to investigate those rotations which bring the vertices and sides into vertices and sides. One such rotation is illustrated in Fig. 4-2.

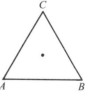 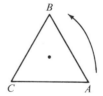

Initial position   Position after rotation

Figure 4-2

This shows the initial position of the triangle and the position after a rotation of 120° has been made in the counterclockwise direction. This same position would be

132   Groups

reached by a rotation of 480° also, since 480 = 360 + 120. The final position of the triangle is our main concern here and for this reason we consider a rotation of 120° in the counterclockwise direction equal to a rotation of 480° in the counterclockwise direction. In fact, *any two rotations having the same effect on the triangle will be equal.*

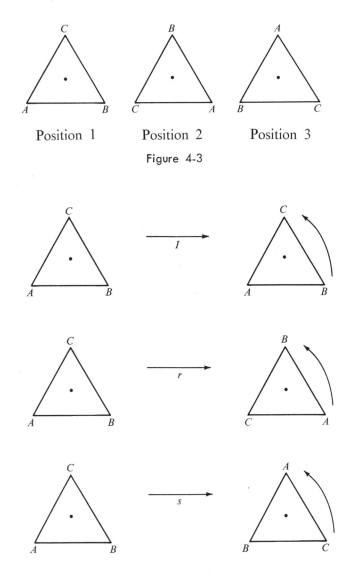

Position 1   Position 2   Position 3

Figure 4-3

Figure 4-4

Groups   133

There are only three positions of the triangle resulting from these rotations (see Fig. 4-3). These positions may be obtained as the result of rotations about 0 as follows:

(a) Position 1 may be obtained as a result of *a rotation in the counterclockwise direction through* 0°.

(b) Position 2 may be obtained as a result of *a rotation in the counterclockwise direction through* 120°.

(c) Position 3 may be obtained as a result of *a rotation in the counterclockwise direction through* 240°.

Let us denote these rotations by I, r, s respectively. The effect of each of these on the initial position of the triangle is illustrated in Fig. 4-4.

A binary operation B may be defined on the set $\{I, r, s\}$ by taking xBy to mean that rotation x is to be followed by rotation y. For example, rBr means that the rotation r is to be applied to the triangle twice, as seen in Fig. 4-5. The result of applying r twice is seen to be the same as applying the rotation s to the triangle. Hence, we have

$$rBr = s$$

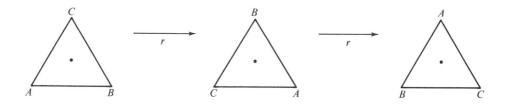

Figure 4-5

In the same way that we have just computed $rBr = s$, we may complete the following table for B.

B	I	r	s
I	I	r	s
r	r	s	I
s	s	I	r

The pair $(\{I, r, s\}, B)$ is a group with I as the identity. The inverse of r is s and r is the inverse of s. The associativity of B will not be shown here; this will be done later.

134 Groups

This is called the group of *congruence motions* of an equilateral triangle in its plane; it is one example of a very important class of groups known as *groups of congruence motions of regular polygons*.

Exercises

(A) In Exercises 1–4, a set G and mappings B are given. Supply the information requested about these.

1. $G = \{1, 2, 3\}$

B	1	2	3
1	2	3	1
2	2	1	3
3	1	2	3

(a) Is B a binary operation on G? _____ Yes

(b) $1B1 =$ _____ ; is 1 an identity element? _____ 2 No

(c) $2B2 =$ _____ ; is 2 an identity element? _____ 1 No

(d) $3B1 =$ _____, $3B2 =$ _____, $3B3 =$ _____ 1, 2, 3
     $1B3 =$ _____, $1B2 =$ _____ 1, 3
     Is 3 an identity element? _____ No

(e) $1B(2B1) = 1B$ _____ = _____ ; 2, 3
     $(1B2)B1 =$ _____ $B1 =$ _____ 3, 1
     Is B associative? _____ No

(f) Is it appropriate to talk about inverses with respect to B? _____ No
     Why? _____ It is not appropriate to talk of inverses without an identity.

Groups   135

2. $G = \{2x \mid x \text{ is an integer}\}$, B is ordinary addition, $+$.

Yes   (a) Is it true that $+: G \times G \longrightarrow G$?

Yes   (b) Is $+$ associative on G?

Yes   (c) Is there an identity for $+$?

$0 = 2 \cdot 0$ (d) What is it?

(e) If $2x \in G$, what is the inverse of $2x$?

$2(-x) = -2x$

Yes   (f) Is the pair $(G, +)$ a group?

3. Let $X = \{1, 2, 3.\}$
   (a) Form all the subsets of X.

$\emptyset, \{1\}, \{2\}, \{3\}, \{1, 2\}, \{1, 3\}, \{2, 3\},$
$\{1, 2, 3\}$

(b) Use the binary operation $\cap$ and complete the table below.

$\cap$	$\emptyset$	$\{1\}$	$\{2\}$	$\{3\}$	$\{1, 2\}$	$\{1, 3\}$	$\{2, 3\}$	$\{1, 2, 3\}$
$\emptyset$								
$\{1\}$								
$\{2\}$								
$\{3\}$								
$\{1, 2\}$								
$\{1, 3\}$								
$\{2, 3\}$								
$\{1, 2, 3\}$								

136 Groups

∩	∅	{1}	{2}	{3}	{1, 2}	{1, 3}	{2, 3}	{1, 2, 3}
∅	∅	∅	∅	∅	∅	∅	∅	∅
{1}	∅	{1}	∅	∅	{1}	{1}	∅	{1}
{2}	∅	∅	{2}	∅	{2}	∅	{2}	{2}
{3}	∅	∅	∅	{3}	∅	{3}	{3}	{3}
{1, 2}	∅	{1}	{2}	∅	{1, 2}	{1}	{2}	{1, 2}
{1, 3}	∅	{1}	∅	{3}	{1}	{1, 3}	{3}	{1, 3}
{2, 3}	∅	∅	{2}	{3}	{2}	{3}	{2, 3}	{2, 3}
{1, 2, 3}	∅	{1}	{2}	{3}	{1, 2}	{1, 3}	{2, 3}	{1, 2, 3}

(c) Does ∩ have an identity element in the set $\{\emptyset, \{1\}, \{2\}, \{3\}, \{1, 2\}, \{1, 3\}, \{2, 3\}, \{1, 2, 3\}\}$?

Yes

What is it? {1, 2, 3}

(d) Does {1} have an inverse with respect to ∩?

No

(e) Do any of the elements of the set $\{\emptyset, \{1\}, \{2\}, \{3\}, \{1, 2\}, \{1, 3\}, \{2, 3\}, \{1, 2, 3\}\}$ have inverses with respect to ∩?

Yes

Which ones have inverses? {1, 2, 3}

(f) Is the pair $(G, \cap)$ a group, where $G = \{\emptyset, \{1\}, \{2\}, \{3\}, \{1, 2\}, \{1, 3\}, \{2, 3\}, \{1, 2, 3\}\}$?

No

4. Let G be the set of Problem 3 above and define a binary operation B on G as follows:
$$xBy = (x \cap y) \cup (y \cap x)$$
(a) Construct a table for B:

Groups 137

B	∅	{1}	{2}	{3}	{1, 2}	{1, 3}	{2, 3}	{1, 2, 3}
∅								
{1}								
{2}								
{3}								
{1, 2}								
{1, 3}								
{2, 3}								
{1, 2, 3}								

B	∅	{1}	{2}	{3}	{1, 2}	{1, 3}	{2, 3}	{1, 2, 3}
∅	∅	{1}	{2}	{3}	{1, 2}	{1, 3}	{2, 3}	{1, 2, 3}
{1}	{1}	∅	{1, 2}	{1, 3}	{2}	{3}	{1, 2, 3}	{2, 3}
{2}	{2}	{1, 2}	∅	{2, 3}	{1}	{1, 2, 3}	{3}	{1, 3}
{3}	{3}	{1, 3}	{2, 3}	∅	{1, 2, 3}	{1}	{2}	{1, 2}
{1, 2}	{1, 2}	{2}	{1}	{1, 2, 3}	∅	{2, 3}	{1, 3}	{3}
{1, 3}	{1, 3}	{3}	{1, 2, 3}	{1}	{2, 3}	∅	{1, 2}	{2}
{2, 3}	{2, 3}	{1, 2, 3}	{3}	{2}	{1, 3}	{1, 2}	∅	{1}
{1, 2, 3}	{1, 2, 3}	{2, 3}	{1, 3}	{1, 2}	{3}	{2}	{1}	∅

Yes, ∅

{1}

{1, 2}

Yes

(b) Does B have an identity element in the set G? _____. What is it? _____
(c) What is the inverse of {1}? _____
(d) What is the inverse of {1, 2}? _____
(e) If B is associative, is (G, B) a group? _____

5. Consider the rotations of the square below about the center 0 which bring the vertices and sides into vertices and sides. As before, two rotations will be equal if they have the same effect on the square. There are four positions which the square can take; these are shown below.

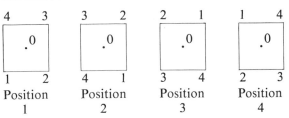

These positions may be obtained by rotation from position 1 as shown below:

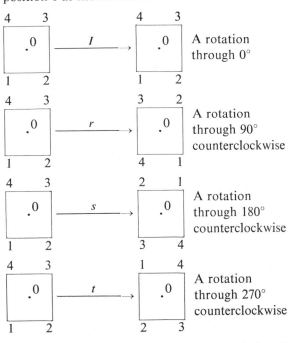

On the set $\{I, r, s, t\}$ define a binary operation B by taking xBy to mean that the rotation x is fol-

Groups   139

lowed by the rotation y. Complete the table below for B.

B	I	r	s	t
I				
r				
s				
t				

(a) Is there an identity in the set $\{I, r, s, t\}$ with respect to B? _____

What is it? _____

(b) Does I have an inverse? _____

What is it? _____

(c) Does r have an inverse? _____

What is it? _____

(d) Does s have an inverse? _____

What is it? _____

(e) Does t have an inverse? _____

What is it? _____

(f) To determine if B is associative we may check all possible equations $(xBy)Bz = xB(yBz)$ to see if these are all true. Since the x, y, z must be members of the set $\{I, r, s, t\}$, in how many ways may each of the x, y, z be chosen? _____

Therefore, we need to check how many different equations? _____

(g) If B is associative, is $(\{I, r, s, t\}, B)$ a group? _____

B	I	r	s	t
I	I	r	s	t
r	r	s	t	I
s	s	t	I	r
t	t	I	r	s

Yes
I
Yes
I
Yes
t
Yes
s
Yes
r

4

$4 \cdot 4 \cdot 4 = 64$

Yes

140 Groups

6. Construct all the 1-1 mappings of the set $\{1, 2, 3\}$ onto itself and make a table for the binary operation B which is the product of mappings. Show that B has an identity and that each element has an inverse with respect to B.

7. In each of the following, decide if (G, B) is a group. If you decide that it is not a group, give a reason for your decision. If (G, B) is a group, give the identity element and the inverse of each element.
   (a) G is the set Z of all integers and B is ordinary addition for Z.
   (b) G is the set Z of all integers and B is ordinary subtraction for Z.
   (c) G is the set Z of all integers and B is ordinary multiplication for Z.
   (d) G is the set of all positive integers and B is addition.
   (e) G is the set of rational numbers (numbers of the form a/b, where $a, b \in Z$ and $b \neq 0$) and B is addition.
   (f) G is the set of rational numbers and B is subtraction.
   (g) G is the set of rational numbers and B is multiplication.
   (h) G is the set of non-zero rational numbers and B is multiplication.

4.2 The Groups $(Z_m, \oplus)$ and $(Z_p^*, \otimes)$

We have studied "congruence, modulo m," where m is an integer such that $m > 1$, in Section 3.5. In Theorem 14 (see p. 121) it was shown that "congruence, modulo m" is an equivalence relation on Z. This means that there is a partition of Z associated with this equivalence relation, "congruence, modulo m" (see Theorem 5, p. 53). We wish to study this partition. Let us first investigate the special case $m = 5$. By Theorem 5, the partition of Z in question is $\{\bar{a} \mid a \in Z\}$. Recall that $\bar{a}$ is defined as follows:
$$\bar{a} = \{x \mid aRx, x \in Z\}$$
where R is an equivalence relation. In this case R is "congruence, mod 5." Therefore,
$$\bar{a} = \{x \mid a \equiv x \ (5), x \in Z\}$$
For example,
$$\bar{2} = \{x \mid 2 \equiv x \ (5), x \in Z\}$$
$$= \{2, 7, 12, 17, \ldots, -3, -8, -13, -18, \ldots\}$$
According to the lemma on p. 53,
$$\bar{a} = \bar{b} \text{ if and only if } aRb$$
for any equivalence R. Here we have
$$\bar{a} = \bar{b} \text{ if and only if } a \equiv b \ (5)$$

Using this result we may show that
$$Z = \bar{0} \cup \bar{1} \cup \bar{2} \cup \bar{3} \cup \bar{4}$$
Another way of saying this is that each integer is in one of the classes $\bar{0}, \bar{1}, \bar{2}, \bar{3}, \bar{4}$. To show this, let a be any integer. Then
$$a = 5q + r, \ 0 \leq r < 5$$
for some integers q, r. The integer a may then be written in one of the following ways:

$$a = 5q \text{ (for } r = 0)$$
$$a = 5q + 1 \text{ (for } r = 1)$$
$$a = 5q + 2 \text{ (for } r = 2)$$
$$a = 5q + 3 \text{ (for } r = 3)$$
$$a = 5q + 4 \text{ (for } r = 4)$$

These equations lead to

$$a - 0 = 5q \text{ or } a \equiv 0 \ (5)$$
$$a - 1 = 5q \text{ or } a \equiv 1 \ (5)$$
$$a - 2 = 5q \text{ or } a \equiv 2 \ (5)$$
$$a - 3 = 5q \text{ or } a \equiv 3 \ (5)$$
$$a - 4 = 5q \text{ or } a \equiv 4 \ (5)$$

Since $\bar{a} = \bar{b}$ if and only if $a \equiv b \ (5)$,

$$\bar{a} = \bar{0} \text{ if } a \equiv 0 \ (5)$$
$$\bar{a} = \bar{1} \text{ if } a \equiv 1 \ (5)$$
$$\bar{a} = \bar{2} \text{ if } a \equiv 2 \ (5)$$
$$\bar{a} = \bar{3} \text{ if } a \equiv 3 \ (5)$$
$$\bar{a} = \bar{4} \text{ if } a \equiv 4 \ (5)$$

But we know that $a \in \bar{a}$ for each a. Hence, $a \in \bar{0}, a \in \bar{1}, a \in \bar{2}, a \in \bar{3}$ or $a \in \bar{4}$.

We denote $\{\bar{0}, \bar{1}, \bar{2}, \bar{3}, \bar{4}\}$ by Z_5, and generally
$$Z_m = \{\bar{0}, \bar{1}, \bar{2}, \bar{3}, \ldots, \overline{(m-1)}\}$$
Just as we have shown above that each integer a is in one of the classes $\bar{0}, \bar{1}, \bar{2}, \bar{3}, \bar{4}$ where the equivalence relation is "congruence, mod 5," we may also show that each integer is a member of one of the classes $\bar{0}, \bar{1}, \bar{2}, \bar{3}, \ldots, \overline{(m-1)}$ when the equivalence relation is "congruence, mod m."

The sets Z_m are important for us here because they lead to some of the most important examples of groups. We obtain these groups by defining an appropriate binary operation on the set Z_m. This binary operation is closely related to addition of integers and will be denoted by $\oplus$. Thus $\oplus : Z_m \times Z_m \to Z_m$ and is defined as follows:
$$\oplus : (\bar{a}, \bar{b}) \longrightarrow \overline{(a+b)} \quad \text{or} \quad \bar{a} \oplus \bar{b} = \overline{(a+b)}$$
where $+$ is ordinary addition of integers. We have implied that this defines a binary

operation on Z_m, but this is premature. We must actually show that $\oplus$ is a mapping of $Z_m \times Z_m$ into Z_m. If $\bar{a} = \bar{c}$ and $\bar{b} = \bar{d}$, then $(\bar{a}, \bar{b}) = (\bar{c}, \bar{d})$ and since

$$\oplus : (\bar{a}, \bar{b}) \longrightarrow \overline{(a + b)}$$

and

$$\oplus : (\bar{c}, \bar{d}) \longrightarrow \overline{(c + d)}$$

we must show that $\overline{(a + b)} = \overline{(c + d)}$. But if $\bar{a} = \bar{c}$ and $\bar{b} = \bar{d}$ it follows that $a \equiv c(m)$ and $b \equiv d\ (m)$ and hence $a + b \equiv c + d\ (m)$. Then $\overline{(a + b)} = \overline{(c + d)}$.

Consider $Z_5 = \{\bar{0}, \bar{1}, \bar{2}, \bar{3}, \bar{4}\}$. By our definition we have, for example,

$$\bar{2} \oplus \bar{4} = \overline{(2 + 4)} = \bar{6}$$

But since $6 \equiv 1\ (5)$, $\bar{6} = \bar{1}$ and

$$\bar{2} \oplus \bar{4} = \bar{1}$$

Computing in this way, we may complete the table below for $\oplus$.

$\oplus$	$\bar{0}$	$\bar{1}$	$\bar{2}$	$\bar{3}$	$\bar{4}$
$\bar{0}$	$\bar{0}$	$\bar{1}$	$\bar{2}$	$\bar{3}$	$\bar{4}$
$\bar{1}$	$\bar{1}$	$\bar{2}$	$\bar{3}$	$\bar{4}$	$\bar{0}$
$\bar{2}$	$\bar{2}$	$\bar{3}$	$\bar{4}$	$\bar{0}$	$\bar{1}$
$\bar{3}$	$\bar{3}$	$\bar{4}$	$\bar{0}$	$\bar{1}$	$\bar{2}$
$\bar{4}$	$\bar{4}$	$\bar{0}$	$\bar{1}$	$\bar{2}$	$\bar{3}$

Theorem 20. $(Z_m, \oplus)$ is a group.

Proof. We have just shown that $\oplus$ is a binary operation on Z_m. $\bar{0}$ is the identity element since for all $\bar{a} \in Z_m$

$$\bar{a} \oplus \bar{0} = \overline{a + 0} = \bar{a}$$

and

$$\bar{0} \oplus \bar{a} = \overline{0 + a} = \bar{a}$$

Also

$$(\bar{a} \oplus \bar{b}) \oplus \bar{c} = \overline{(a + b)} \oplus \bar{c} = \overline{(a + b) + c}$$

and $\bar{a} \oplus (\bar{b} \oplus \bar{c}) = \bar{a} \oplus \overline{(b + c)} = \overline{a + (b + c)}$. Since $(a + b) + c = a + (b + c)$, it follows that

$$(\bar{a} \oplus \bar{b}) \oplus \bar{c} = \bar{a} \oplus (\bar{b} \oplus \bar{c})$$

which shows that $\oplus$ is associative. For $\bar{a} \in Z_m$, $\overline{(-a)} \in Z_m$ is its inverse:

$$\bar{a} \oplus \overline{(-a)} = \overline{a + (-a)} = \bar{0}$$

and
$$(\overline{-a}) \oplus \bar{a} = \overline{(-a) + a} = \bar{0}$$
This completes the proof that $(Z_m, +)$ is a group.

In a manner similar to the way in which $\oplus$ was defined for Z_m we may also define another binary operation $\otimes$ on Z_m as follows: for all $\bar{a}, \bar{b} \in Z_m$
$$\otimes : (\bar{a}, \bar{b}) \longrightarrow \overline{(a \cdot b)} \quad \text{or} \quad \bar{a} \otimes \bar{b} = \overline{(a \cdot b)}$$
Again it is necessary to show that this is actually a binary operation on Z_m. If $\bar{a} = \bar{b}$ and $\bar{c} = \bar{d}$, then $(\bar{a}, \bar{c}) = (\bar{b}, \bar{d})$. By definition
$$\otimes : (\bar{a}, \bar{c}) \longrightarrow \overline{(a \cdot c)}$$
and
$$\otimes : (\bar{b}, \bar{d}) \longrightarrow \overline{(b \cdot d)}$$
Hence, for $\otimes$ to be a mapping we must show that $\overline{(a \cdot c)} = \overline{(b \cdot d)}$. But if $\bar{a} = \bar{b}$ and $\bar{c} = \bar{d}$, then $a \equiv b \ (m)$ and $c \equiv d \ (m)$ from which it follows that $a \cdot c \equiv b \cdot d \ (m)$ and finally $\overline{(a \cdot c)} = \overline{(b \cdot d)}$. The class $\bar{1}$ is the identity element here since for any $\bar{a} \in Z_m$
$$\bar{1} \otimes \bar{a} = \overline{(1 \cdot a)} = \bar{a}$$
and
$$\bar{a} \otimes \bar{1} = \overline{(a \cdot 1)} = \bar{a}$$
The binary operation $\otimes$ is easily shown to be associative. It would not be unreasonable at this point to expect that $(Z_m, \otimes)$ is a group. This, however, is not the case. One reason for this is that $\bar{0} \in Z_m$ and for $\bar{x} \in Z_m$
$$\bar{x} \otimes \bar{0} = \overline{(x \cdot 0)} = \bar{0}$$
This shows that $\bar{0}$ has no inverse. A natural way to remedy this might appear to be the elimination of $\bar{0}$ from consideration. Define $Z_m^* = Z_m \cap \{\bar{0}\}$. Now we consider $(Z_m^*, \otimes)$. Suppose that m is not prime. Then $m = a \cdot b$ for some integers a, b such that
$$1 < a < m \quad \text{and} \quad 1 < b < m$$
Computing $\bar{a} \otimes \bar{b}$ we have
$$\bar{a} \otimes \bar{b} = \overline{a \cdot b} = \bar{m} = \bar{0}$$
since $m \equiv 0 \ (m)$. This shows that $\otimes$ is *not* a binary operation on Z_m^* since $\bar{0} \notin Z_m^*$. Fortunately, this situation cannot arise if m is prime. Suppose to the contrary that m is prime and that for $\bar{a}, \bar{b} \in Z_m^*$ we have $\bar{a} \otimes \bar{b} = \bar{0}$. Then by definition of $\otimes$, $\overline{a \cdot b} = \bar{0}$ and hence $a \cdot b \equiv 0 \ (m)$. Therefore, $m \,|\, a \cdot b$ and since m is prime we conclude that $m \,|\, a$ or $m \,|\, b$ (see Theorem 13, p. 117). If $m \,|\, a$, then $a \equiv 0 \ (m)$ so that $\bar{a} = \bar{0}$ and $\bar{a} \notin Z_m^*$ which is a contradiction. The same contradiction occurs if $m \,|\, b$. Thus we see that $\otimes : Z_m^* \times Z_m^* \to Z_m^*$ *provided* m is prime. Let $\bar{a} \in Z_m^*$. Then we may assume that $a < m$ so that if m is prime GCD $(a, m) = 1$. There are, therefore, integers x, y such

Groups

that $1 = ax + my$. Therefore, $1 - ax = my$, $m\,|\,(1 - ax)$ so that
$$ax \equiv 1 \ (m)$$
Then $\overline{ax} = \overline{1}$ or $\bar{a} \otimes \bar{x} = \bar{1}$. This shows that there is an element $\bar{x} \in Z_m^*$ which is the inverse of $\bar{a}$. Since $\otimes$ is associative, we conclude that $(Z_m^*, \otimes)$ is a group if and only if m is prime.

Exercises

$10 \equiv 1 \ (3)$

$21 \equiv 0 \ (7)$

$25 \equiv 5 \ (10)$

$-10 \equiv 4 \ (14)$

$-8 \equiv 20 \ (7)$

$\overline{1}$

$\overline{7}$

$\overline{6}$

$\overline{8}$

$\overline{15}$

8. In Z_3, $\overline{10} = \overline{1}$ because _____

9. In Z_7, $\overline{21} = \overline{0}$ because _____

10. In Z_{10}, $\overline{25} = \overline{5}$ because _____

11. In Z_{14}, $\overline{(-10)} = \overline{4}$ because _____

12. In Z_7, $\overline{(-8)} = \overline{(20)}$ because _____

In Exercises 13–17, an equivalence class $\bar{a}$ is given in some Z_m. Find an equivalence class $\bar{r} \in Z_m$ such that $\bar{a} = \bar{r}$ and $0 \leq r < m$.

13. $\overline{10} \in Z_3$. $\overline{10} = $ _____

14. $\overline{18} \in Z_{11}$. $\overline{18} = $ _____

15. $\overline{(-8)} \in Z_{14}$. $\overline{(-8)} = $ _____

16. $\overline{84} \in Z_{19}$. $\overline{84} = $ _____

17. $\overline{111} \in Z_{32}$. $\overline{111} = $ _____

Groups   145

In Exercises 18–27, the final answer to the computations should be some $\bar{r} \in Z_m$ with $0 \leq r < m$.

$\overline{13}, \bar{3}$

18. In $(Z_{10}, \oplus)$, $\bar{7} \oplus \bar{6} = $ _____ = _____

$\overline{31}, \overline{10}$

19. In $(Z_{21}, \oplus)$, $\overline{12} \oplus \overline{19} = $ _____ = _____

$\overline{18}, \bar{0}$

20. In $(Z_{18}, \oplus)$, $\bar{5} \oplus \overline{13} = $ _____ = _____

$\overline{184}, \overline{84}$

21. In $(Z_{100}, \oplus)$, $\overline{85} \oplus \overline{99} = $ _____ = _____

$(\overline{-105}), \overline{176}$

22. In $(Z_{281}, \oplus)$, $(\overline{-10}) \oplus (\overline{-95}) = $ _____ = _____

$\overline{16}, \bar{5}$

23. In $(Z^*_{11}, \otimes)$, $\bar{2} \otimes \bar{8} = $ _____ = _____

$\overline{100}, \bar{1}$

24. In $(Z^*_{11}, \otimes)$, $\overline{10} \otimes \overline{10} = $ _____ = _____

$\overline{52}, \bar{6}$

25. In $(Z^*_{23}, \otimes)$, $\overline{13} \otimes \bar{4} = $ _____ = _____

$\overline{195}, \bar{9}$

26. In $(Z^*_{31}, \otimes)$, $\overline{15} \otimes \overline{13} = $ _____ = _____

$\overline{468}, \bar{3}$

27. In $(Z^*_{31}, \otimes)$, $\overline{26} \otimes \overline{18} = $ _____ = _____

28. Complete the tables below for $(Z_2, \oplus)$ and $(Z^*_2, \otimes)$

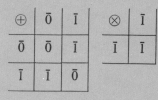

$\oplus$	$\bar{0}$	$\bar{1}$
$\bar{0}$		
$\bar{1}$		

$\otimes$	$\bar{1}$
$\bar{1}$	

$\bar{0}$
$\bar{1}$

(a) What is the identity for $(Z_2, \oplus)$? _____
(b) What is the identity for $(Z^*_2, \otimes)$? _____

(c) What is the inverse of $\bar{0}$ in $(Z_2, \oplus)$? _____

(d) What is the inverse of $\bar{1}$ in $(Z_2, \oplus)$? _____

(e) What is the inverse of $\bar{1}$ in $(Z_2^*, \otimes)$? _____

29. Complete the tables below for $(Z_3, \oplus)$ and $(Z_3^*, \otimes)$

$\oplus$	$\bar{0}$	$\bar{1}$	$\bar{2}$
$\bar{0}$			
$\bar{1}$			
$\bar{2}$			

$\otimes$	$\bar{1}$	$\bar{2}$
$\bar{1}$		
$\bar{2}$		

(a) What is the identity for $(Z_3, \oplus)$? _____

(b) What is the identity for $(Z_3^*, \otimes)$? _____

(c) What is the inverse of $\bar{0}$ in $(Z_3, \oplus)$? _____

(d) What is the inverse of $\bar{1}$ in $(Z_3, \oplus)$? _____

(e) What is the inverse of $\bar{2}$ in $(Z_3, \oplus)$? _____

(f) What is the inverse of $\bar{1}$ in $(Z_3^*, \otimes)$? _____

(g) What is the inverse of $\bar{2}$ in $(Z_3^*, \otimes)$? _____

30. Complete the table below for $(Z_4, \oplus)$.

$\oplus$	$\bar{0}$	$\bar{1}$	$\bar{2}$	$\bar{3}$
$\bar{0}$				
$\bar{1}$				
$\bar{2}$				
$\bar{3}$				

Groups   147

31. Complete the table below for $(Z_6, \oplus)$.

$\oplus$	$\bar{0}$	$\bar{1}$	$\bar{2}$	$\bar{3}$	$\bar{4}$	$\bar{5}$
$\bar{0}$	$\bar{0}$	$\bar{1}$	$\bar{2}$	$\bar{3}$	$\bar{4}$	$\bar{5}$
$\bar{1}$	$\bar{1}$	$\bar{2}$	$\bar{3}$	$\bar{4}$	$\bar{5}$	$\bar{0}$
$\bar{2}$	$\bar{2}$	$\bar{3}$	$\bar{4}$	$\bar{5}$	$\bar{0}$	$\bar{1}$
$\bar{3}$	$\bar{3}$	$\bar{4}$	$\bar{5}$	$\bar{0}$	$\bar{1}$	$\bar{2}$
$\bar{4}$	$\bar{4}$	$\bar{5}$	$\bar{0}$	$\bar{1}$	$\bar{2}$	$\bar{3}$
$\bar{5}$	$\bar{5}$	$\bar{0}$	$\bar{1}$	$\bar{2}$	$\bar{3}$	$\bar{4}$

(a) What is the identity element for $(Z_6, \oplus)$? $\bar{0}$

(b) What is the inverse of $\bar{0}$ in $(Z_6, \oplus)$? $\bar{0}$
(c) What is the inverse of $\bar{1}$ in $(Z_6, \oplus)$? $\bar{5}$
(d) What is the inverse of $\bar{2}$ in $(Z_6, \oplus)$? $\bar{4}$
(e) What is the inverse of $\bar{3}$ in $(Z_6, \oplus)$? $\bar{3}$
(f) What is the inverse of $\bar{4}$ in $(Z_6, \oplus)$? $\bar{2}$
(g) What is the inverse of $\bar{5}$ in $(Z_6, \oplus)$? $\bar{1}$
(h) Is $(Z_6^*, \otimes)$ a group?   No   Why?   Because 6 is not prime.

Groups

$\oplus$	$\bar{0}$	$\bar{1}$	$\bar{2}$	$\bar{3}$	$\bar{4}$	$\bar{5}$	$\bar{6}$
$\bar{0}$	$\bar{0}$	$\bar{1}$	$\bar{2}$	$\bar{3}$	$\bar{4}$	$\bar{5}$	$\bar{6}$
$\bar{1}$	$\bar{1}$	$\bar{2}$	$\bar{3}$	$\bar{4}$	$\bar{5}$	$\bar{6}$	$\bar{0}$
$\bar{2}$	$\bar{2}$	$\bar{3}$	$\bar{4}$	$\bar{5}$	$\bar{6}$	$\bar{0}$	$\bar{1}$
$\bar{3}$	$\bar{3}$	$\bar{4}$	$\bar{5}$	$\bar{6}$	$\bar{0}$	$\bar{1}$	$\bar{2}$
$\bar{4}$	$\bar{4}$	$\bar{5}$	$\bar{6}$	$\bar{0}$	$\bar{1}$	$\bar{2}$	$\bar{3}$
$\bar{5}$	$\bar{5}$	$\bar{6}$	$\bar{0}$	$\bar{1}$	$\bar{2}$	$\bar{3}$	$\bar{4}$
$\bar{6}$	$\bar{6}$	$\bar{0}$	$\bar{1}$	$\bar{2}$	$\bar{3}$	$\bar{4}$	$\bar{5}$

$\otimes$	$\bar{1}$	$\bar{2}$	$\bar{3}$	$\bar{4}$	$\bar{5}$	$\bar{6}$
$\bar{1}$	$\bar{1}$	$\bar{2}$	$\bar{3}$	$\bar{4}$	$\bar{5}$	$\bar{6}$
$\bar{2}$	$\bar{2}$	$\bar{4}$	$\bar{6}$	$\bar{1}$	$\bar{3}$	$\bar{5}$
$\bar{3}$	$\bar{3}$	$\bar{6}$	$\bar{2}$	$\bar{5}$	$\bar{1}$	$\bar{4}$
$\bar{4}$	$\bar{4}$	$\bar{1}$	$\bar{5}$	$\bar{2}$	$\bar{6}$	$\bar{3}$
$\bar{5}$	$\bar{5}$	$\bar{3}$	$\bar{1}$	$\bar{6}$	$\bar{4}$	$\bar{2}$
$\bar{6}$	$\bar{6}$	$\bar{5}$	$\bar{4}$	$\bar{3}$	$\bar{2}$	$\bar{1}$

$\bar{4}$

$\overline{12}$

32. Complete the tables below for $(Z_7, \oplus)$ and $(Z_7^*, \otimes)$.

$\oplus$	$\bar{0}$	$\bar{1}$	$\bar{2}$	$\bar{3}$	$\bar{4}$	$\bar{5}$	$\bar{6}$
$\bar{0}$							
$\bar{1}$							
$\bar{2}$							
$\bar{3}$							
$\bar{4}$							
$\bar{5}$							
$\bar{6}$							

$\otimes$	$\bar{1}$	$\bar{2}$	$\bar{3}$	$\bar{4}$	$\bar{5}$	$\bar{6}$
$\bar{1}$						
$\bar{2}$						
$\bar{3}$						
$\bar{4}$						
$\bar{5}$						
$\bar{6}$						

33. Show that the associative and commutative laws hold in $(Z_m^*, \otimes)$.

34. In Z_5 find an x such that $\bar{3} \oplus \bar{x} = \bar{2}$, i.e., solve the equation $\bar{3} \oplus \bar{x} = \bar{2}$. _____

35. In Z_{25} solve the equation $\bar{y} \oplus \overline{18} = \bar{5}$. _____

36. In Z_{13} solve the equation $\bar{z} \otimes \bar{5} = \bar{4}$. _____

37. In Z_{31} solve the equation $\overline{11} \otimes \bar{w} = \overline{15}$. _____

4.3 Some Elementary Properties of Groups

There are a number of elementary deductions which may be obtained from the definition of a group. For example, if (G, B) is a group with binary operation B, the associative law states that

$$(xBy)Bz = xB(yBz)$$

for all $x, y, z \in G$. Because of this law it is customary to omit parentheses and just write $xByBz$. If four elements are involved, say x, y, z, w, we may show that the element $xB(yB(zBw))$ is the same as $(xBy)B(zBw)$ or $((xBy)Bz)w$. Thus we simply write $xByBzBw$ for this element. Generally, it is possible to show that parentheses may be omitted no matter how many group elements are involved.*

A notational practice that is commonly used is the usual multiplication notation to stand for any binary operation B. For example, instead of writing xBy, one writes $x \cdot y$ or simply xy. The advantage of doing this is that it is much easier to write xy or $x \cdot y$ instead of xBy. *One should not assume that ordinary multiplication is being considered unless this is specifically stated.* In this notation the associative law is

$$(xy)z = x(yz)$$

for all $x, y, z \in G$.

The definition of a group $(G, \cdot)$ states that there is an identity element e and that each element $g \in G$ has an inverse $h \in G$ and that these elements are related to each other by the equations

$$x \cdot e = x, \, e \cdot x = x \text{ for all } x \in G$$
$$g \cdot h = e, \, h \cdot g = e$$

The definition does not state that there is only one identity element nor does it state that an element g can have only one inverse. This is the case, however, as the following theorem shows.

Theorem 21. Let $(G, \cdot)$ be a group. (a) The identity element e of $(G, \cdot)$ is unique. (b) Each element $g \in G$ has only one inverse in G.

Proof. (a) Suppose that f is an identity element of $(G, \cdot)$ as well as e. Then $x \cdot e = x = e \cdot x$ for all $x \in G$ and $x \cdot f = x = f \cdot x$ for all $x \in G$. Therefore, in particular,

---
*See Baumslag, B., and B. Chandler, *Group Theory* (New York: McGraw-Hill, 1968), p. 39.

$f \cdot e = f$ by the first equations and $e = f \cdot e$ by the second equations. Hence, $e = f$ proving part (a).

(b) Suppose that x and y are both inverses of g, i.e.,
$$g \cdot x = x \cdot g = e \text{ and } g \cdot y = y \cdot g = e$$
Then
$$\begin{aligned} x &= x \cdot e \\ &= x \cdot (g \cdot y) & \text{substituting } g \cdot y \text{ for } e \\ &= (x \cdot g) \cdot y & \text{associative law} \\ &= e \cdot y & \text{substituting } e \text{ for } x \cdot g \\ &= y \end{aligned}$$

Therefore, g has only one inverse. (The unique inverse of g will be denoted by g^{-1}.)

Corollary. If a, b are elements of a group $(G, \cdot)$, then $(ab)^{-1} = b^{-1} \cdot a^{-1}$.

Proof. The element $(ab)^{-1}$ is the inverse of ab. But $b^{-1}a^{-1}$ is also an inverse of ab since
$$\begin{aligned} (ab) \cdot (b^{-1}a^{-1}) &= a(bb^{-1})a^{-1} \\ &= a \cdot e \cdot a^{-1} \\ &= a \cdot a^{-1} = e \end{aligned}$$

By the theorem, ab has only one inverse. Therefore, $(ab)^{-1} = b^{-1}a^{-1}$.

Theorem 22. Let $(G, \cdot)$ be a group and let $a, b \in G$. The equations
$$ax = b \text{ and } ya = b$$
have unique solutions in G.

Proof. A solution of the equation $ax = b$ is $x = a^{-1}b$:
$$\begin{aligned} ax &= a(a^{-1}b) \\ &= (aa^{-1})b \\ &= e \cdot b = b \end{aligned}$$

Also $y = ba^{-1}$ is a solution of the equation $ya = b$:
$$\begin{aligned} ya &= (ba^{-1})a \\ &= b(a^{-1}a) \\ &= b \cdot e = b \end{aligned}$$

This shows that the two equations have solutions. Now suppose that w, z are both solutions for the equation $ax = b$, i.e., suppose $aw = b$ and $az = b$. Then, since aw and az are both b,
$$aw = az$$

Hence,
$$a^{-1}(aw) = a^{-1}(az)$$
$$(a^{-1}a)w = (a^{-1}a)z$$
$$ew = ez$$
$$w = z$$

Therefore, any two solutions of the equation $ax = b$ are equal so that the solution $a^{-1}b$ is unique. In like manner it is shown that the solution ba^{-1} for the equation $ya = b$ is unique.

If a is some particular element in a group $(G, \cdot)$ we may define a mapping of G into G which is associated with a by mapping any element $x \in G$ onto the element $x \cdot a$. If we call this mapping R_a (short for right multiplication by a), we have

$$R_a: G \longrightarrow G$$

such that

$$R_a: x \longrightarrow x \cdot a \text{ for all } x \in G$$

In a similar way we define

$$L_a: G \longrightarrow G$$

such that

$$L_a: x \longrightarrow a \cdot x \text{ for all } x \in G$$

Theorem 23. Let $(G, \cdot)$ be a group and let a be any element in G. Then both R_a and L_a are 1–1 mappings of G onto G.

Proof. The domain of both R_a and L_a is G by definition. To show that R_a is a mapping we must show—by definition of mapping—that if

$$R_a: x \longrightarrow x \cdot a$$

and

$$R_a: y \longrightarrow y \cdot a$$

and $x = y$, then $x \cdot a = y \cdot a$. But this is immediate since if $x = y$, $x \cdot a = y \cdot a$ because $\cdot$ is a binary operation on G. Also $x \cdot a \in G$ for all $x \in G$. Hence, R_a is a mapping of G into G. To show that R_a is 1–1, suppose that $x \cdot a = y \cdot a$. Then

$$(x \cdot a) \cdot a^{-1} = (y \cdot a) \cdot a^{-1}$$
$$x \cdot e = y \cdot e$$
$$x = y$$

Therefore, by definition R_a is 1–1. Finally, to show that R_a is a mapping of G *onto* G, it must be shown that if b is any element of G, then there is an element $x \in G$ such that

$$R_a: x \longrightarrow b$$

By Theorem 22 there is a unique solution for the equation $x \cdot a = b$. Then

$$R_a: \quad x \longrightarrow x \cdot a = b$$

so that R_a is onto. This completes the proof that R_a is a 1-1 mapping of G onto G. The proof for L_a is done in a similar way.

Suppose that $(G, \cdot)$ is a group and suppose that G is a finite set, say $G = \{a_1, a_2, a_3, \ldots, a_n\}$. Then the binary operation on G may be described in a table as we have done before with binary operations (see Problems 117 and 118 of Chapter 2 and examples of Section 4.1). Theorem 23 above has an interpretation in this group table. By this theorem, if $x \neq y$, then $x \cdot a \neq y \cdot a$. Suppose that the elements x, y, a are in the group table as shown:

$\cdot$	a_1	a_2	$\cdots$	a	$\cdots$	a_n
a_1						
a_2						
$\vdots$						
x				$x \cdot a$		
$\vdots$						
y				$y \cdot a$		
$\vdots$						
a_n						

The interpretation is then this: *in the a-column, different rows ($x \neq y$) have different elements ($x \cdot a \neq y \cdot a$).* Also, since R_a is a mapping of G onto G, each element of G occurs in the a-column. Of course, the same conclusions may be drawn about the a-row of the table. We may then say that *in the group table, each element of G occurs exactly once in each row and each column.*

Example 4-6: We have seen that ordinary multiplication on the set $\{1, -1\}$ gives a group. This has the following group table.

·	1	−1
1	1	−1
−1	−1	1

Notice that the elements $1, -1$ occur exactly once in each row and each column. Now we construct R_1, R_{-1}, L_1 and L_{-1}:

$$R_1: \quad 1 \longrightarrow 1 \cdot 1 = 1$$
$$R_1: \quad -1 \longrightarrow -1 \cdot 1 = -1 \quad \text{or } R_1 = \begin{pmatrix} 1 & -1 \\ 1 & -1 \end{pmatrix}$$

$$R_{-1}: \quad 1 \longrightarrow 1 \cdot (-1) = -1$$
$$R_{-1}: \quad -1 \longrightarrow (-1) \cdot (-1) = 1 \quad \text{or } R_{-1} = \begin{pmatrix} 1 & -1 \\ -1 & 1 \end{pmatrix}$$

$$L_1: \quad 1 \longrightarrow 1 \cdot 1 = 1$$
$$L_1: \quad -1 \longrightarrow 1 \cdot (-1) = -1 \quad \text{or } L_1 = \begin{pmatrix} 1 & -1 \\ 1 & -1 \end{pmatrix}$$

$$L_{-1}: \quad 1 \longrightarrow (-1) \cdot 1 = -1$$
$$L_{-1}: \quad -1 \longrightarrow (-1) \cdot (-1) = 1 \quad \text{or } L_{-1} = \begin{pmatrix} 1 & -1 \\ -1 & 1 \end{pmatrix}$$

Example 4-7: Consider the group table below for the group $(Z_5, \oplus)$.

$\oplus$	$\bar{0}$	$\bar{1}$	$\bar{2}$	$\bar{3}$	$\bar{4}$
$\bar{0}$	$\bar{0}$	$\bar{1}$	$\bar{2}$	$\bar{3}$	$\bar{4}$
$\bar{1}$	$\bar{1}$	$\bar{2}$	$\bar{3}$	$\bar{4}$	$\bar{0}$
$\bar{2}$	$\bar{2}$	$\bar{3}$	$\bar{4}$	$\bar{0}$	$\bar{1}$
$\bar{3}$	$\bar{3}$	$\bar{4}$	$\bar{0}$	$\bar{1}$	$\bar{2}$
$\bar{4}$	$\bar{4}$	$\bar{0}$	$\bar{1}$	$\bar{2}$	$\bar{3}$

Observe that each of the elements $\bar{0}, \bar{1}, \bar{2}, \bar{3}, \bar{4}$ occurs exactly once in each row and each column of the table. From the table we compute $R_{\bar{2}}$, $R_{\bar{4}}$, and $L_{\bar{3}}$.

$$R_{\bar{2}}: \quad \bar{0} \longrightarrow \bar{0} \oplus \bar{2} = \bar{2}$$
$$R_{\bar{2}}: \quad \bar{1} \longrightarrow \bar{1} \oplus \bar{2} = \bar{3}$$
$$R_{\bar{2}}: \quad \bar{2} \longrightarrow \bar{2} \oplus \bar{2} = \bar{4} \quad \text{or } R_{\bar{2}} = \begin{pmatrix} \bar{0} & \bar{1} & \bar{2} & \bar{3} & \bar{4} \\ \bar{2} & \bar{3} & \bar{4} & \bar{0} & \bar{1} \end{pmatrix}$$
$$R_{\bar{2}}: \quad \bar{3} \longrightarrow \bar{3} \oplus \bar{2} = \bar{0}$$
$$R_{\bar{2}}: \quad \bar{4} \longrightarrow \bar{4} \oplus \bar{2} = \bar{1}$$

$$R_{\bar{4}}: \quad \bar{0} \longrightarrow \bar{0} \oplus \bar{4} = \bar{4}$$
$$R_{\bar{4}}: \quad \bar{1} \longrightarrow \bar{1} \oplus \bar{4} = \bar{0}$$
$$R_{\bar{4}}: \quad \bar{2} \longrightarrow \bar{2} \oplus \bar{4} = \bar{1} \quad \text{or } R_{\bar{4}} = \begin{pmatrix} \bar{0} & \bar{1} & \bar{2} & \bar{3} & \bar{4} \\ \bar{4} & \bar{0} & \bar{1} & \bar{2} & \bar{3} \end{pmatrix}$$
$$R_{\bar{4}}: \quad \bar{3} \longrightarrow \bar{3} \oplus \bar{4} = \bar{2}$$
$$R_{\bar{4}}: \quad \bar{4} \longrightarrow \bar{4} \oplus \bar{4} = \bar{3}$$

$L_{\bar{3}}: \bar{0} \longrightarrow \bar{3} \oplus \bar{0} = \bar{3}$
$L_{\bar{3}}: \bar{1} \longrightarrow \bar{3} \oplus \bar{1} = \bar{4}$
$L_{\bar{3}}: \bar{2} \longrightarrow \bar{3} \oplus \bar{2} = \bar{0}$ or $L_{\bar{3}} = \begin{pmatrix} \bar{0} & \bar{1} & \bar{2} & \bar{3} & \bar{4} \\ \bar{3} & \bar{4} & \bar{0} & \bar{1} & \bar{2} \end{pmatrix}$
$L_{\bar{3}}: \bar{3} \longrightarrow \bar{3} \oplus \bar{3} = \bar{1}$
$L_{\bar{3}}: \bar{4} \longrightarrow \bar{3} \oplus \bar{4} = \bar{2}$

Exercises

38. Complete the table for $(Z_{\bar{3}}^*, \otimes)$ below and compute R_a and L_a for each $\bar{a} \in Z_{\bar{3}}^*$.

$\otimes$	$\bar{1}$	$\bar{2}$
$\bar{1}$	$\bar{1}$	$\bar{2}$
$\bar{2}$	$\bar{2}$	$\bar{1}$

$\otimes$	$\bar{1}$	$\bar{2}$
$\bar{1}$		
$\bar{2}$		

$\bar{1}$
$\bar{2}, \bar{1}\,\bar{2}$
$\bar{2}$
$\bar{1}, \bar{2}\,\bar{1}$
$\bar{1}$
$\bar{2}, \bar{1}\,\bar{2}$
$\bar{2}$
$\bar{1}, \bar{2}\,\bar{1}$

$R_{\bar{1}}: \bar{1} \longrightarrow$ _____
$R_{\bar{1}}: \bar{2} \longrightarrow$ _____ or $R_{\bar{1}} = \begin{pmatrix} \bar{1} & \bar{2} \\ & \end{pmatrix}$

$R_{\bar{2}}: \bar{1} \longrightarrow$ _____
$R_{\bar{2}}: \bar{2} \longrightarrow$ _____ or $R_{\bar{2}} = \begin{pmatrix} \bar{1} & \bar{2} \\ & \end{pmatrix}$

$L_{\bar{1}}: \bar{1} \longrightarrow$ _____
$L_{\bar{1}}: \bar{2} \longrightarrow$ _____ or $L_{\bar{1}} = \begin{pmatrix} \bar{1} & \bar{2} \\ & \end{pmatrix}$

$L_{\bar{2}}: \bar{1} \longrightarrow$ _____
$L_{\bar{2}}: \bar{2} \longrightarrow$ _____ or $L_{\bar{2}} = \begin{pmatrix} \bar{1} & \bar{2} \\ & \end{pmatrix}$

39. Complete the table for $(Z_{\bar{5}}^*, \otimes)$ below and compute R_a and L_a for each $\bar{a} \in Z_{\bar{5}}^*$.

$\otimes$	$\bar{1}$	$\bar{2}$	$\bar{3}$	$\bar{4}$
$\bar{1}$	$\bar{1}$	$\bar{2}$	$\bar{3}$	$\bar{4}$
$\bar{2}$	$\bar{2}$	$\bar{4}$	$\bar{1}$	$\bar{3}$
$\bar{3}$	$\bar{3}$	$\bar{1}$	$\bar{4}$	$\bar{2}$
$\bar{4}$	$\bar{4}$	$\bar{3}$	$\bar{2}$	$\bar{1}$

$\otimes$	$\bar{1}$	$\bar{2}$	$\bar{3}$	$\bar{4}$
$\bar{1}$				
$\bar{2}$				
$\bar{3}$				
$\bar{4}$				

Groups   155

$\bar{1}\bar{2}\bar{3}\bar{4}$, $\bar{2}\bar{4}\bar{1}\bar{3}$

$\bar{3}\bar{1}\bar{4}\bar{2}$, $\bar{4}\bar{3}\bar{2}\bar{1}$

$\bar{1}\bar{2}\bar{3}\bar{4}$, $\bar{2}\bar{4}\bar{1}\bar{3}$

$\bar{3}\bar{1}\bar{4}\bar{2}$, $\bar{4}\bar{3}\bar{2}\bar{1}$

$R_{\bar{1}} = \begin{pmatrix} \bar{1} & \bar{2} & \bar{3} & \bar{4} \\ & & & \end{pmatrix}$, $R_{\bar{2}} = \begin{pmatrix} \bar{1} & \bar{2} & \bar{3} & \bar{4} \\ & & & \end{pmatrix}$

$R_{\bar{3}} = \begin{pmatrix} \bar{1} & \bar{2} & \bar{3} & \bar{4} \\ & & & \end{pmatrix}$, $R_{\bar{4}} = \begin{pmatrix} \bar{1} & \bar{2} & \bar{3} & \bar{4} \\ & & & \end{pmatrix}$

$L_{\bar{1}} = \begin{pmatrix} \bar{1} & \bar{2} & \bar{3} & \bar{4} \\ & & & \end{pmatrix}$, $L_{\bar{2}} = \begin{pmatrix} \bar{1} & \bar{2} & \bar{3} & \bar{4} \\ & & & \end{pmatrix}$

$L_{\bar{3}} = \begin{pmatrix} \bar{1} & \bar{2} & \bar{3} & \bar{4} \\ & & & \end{pmatrix}$, $L_{\bar{4}} = \begin{pmatrix} \bar{1} & \bar{2} & \bar{3} & \bar{4} \\ & & & \end{pmatrix}$

40. The following is the group table for the six 1–1 mappings of a set of 3 elements onto itself where the binary operation is the product of mappings. Use this table to compute all the R_a and L_a.

·	e	x	y	z	w	v
e	e	x	y	z	w	v
x	x	e	w	v	y	z
y	y	v	e	w	z	x
z	z	w	v	e	x	y
w	w	z	x	y	v	e
v	v	y	z	x	e	w

$e\,x\,y\,z\,w\,v$, $e\,x\,y\,z\,w\,v$

$x\,e\,v\,w\,z\,y$, $x\,e\,w\,v\,y\,z$

$y\,w\,e\,v\,x\,z$, $y\,v\,e\,w\,z\,x$

$z\,v\,w\,e\,y\,x$, $z\,w\,v\,e\,x\,y$

$R_e = \begin{pmatrix} e & x & y & z & w & v \\ & & & & & \end{pmatrix}$ $L_e = \begin{pmatrix} e & x & y & z & w & v \\ & & & & & \end{pmatrix}$

$R_x = \begin{pmatrix} e & x & y & z & w & v \\ & & & & & \end{pmatrix}$ $L_x = \begin{pmatrix} e & x & y & z & w & v \\ & & & & & \end{pmatrix}$

$R_y = \begin{pmatrix} e & x & y & z & w & v \\ & & & & & \end{pmatrix}$ $L_y = \begin{pmatrix} e & x & y & z & w & v \\ & & & & & \end{pmatrix}$

$R_z = \begin{pmatrix} e & x & y & z & w & v \\ & & & & & \end{pmatrix}$ $L_z = \begin{pmatrix} e & x & y & z & w & v \\ & & & & & \end{pmatrix}$

156   Groups

w y z x v e,   w z x y v e

v z x y e w,   v y z x e w

$\bar{4}$
$\bar{3}$
$\bar{4}$
$\bar{1}, \bar{4}$

z
v
z
v
e
z

$R_w = \begin{pmatrix} e & x & y & z & w & v \\ \rule{1cm}{0.4pt} & & & & & \end{pmatrix}$ $L_w = \begin{pmatrix} e & x & y & z & w & v \\ \rule{1cm}{0.4pt} & & & & & \end{pmatrix}$

$R_v = \begin{pmatrix} e & x & y & z & w & v \\ \rule{1cm}{0.4pt} & & & & & \end{pmatrix}$ $L_v = \begin{pmatrix} e & x & y & z & w & v \\ \rule{1cm}{0.4pt} & & & & & \end{pmatrix}$

41. Refer to your group table in Exercise 39 above and solve each of the following equations for x.

   (a) $\bar{2} \otimes x = \bar{3}$. Answer: $x =$ _____
   (b) $\bar{3} \otimes x = \bar{4}$. Answer: $x =$ _____
   (c) $x \otimes \bar{4} = \bar{1}$. Answer: $x =$ _____
   (d) $x \otimes x = \bar{1}$. Answer: $x =$ _____

42. Refer to your group table in Exercise 40 above and solve each of the following equations for u.

   (a) $x \cdot u = v$. Answer: $u =$ _____
   (b) $u \cdot z = x$. Answer: $u =$ _____
   (c) $y \cdot u = w$. Answer: $u =$ _____
   (d) $x \cdot u \cdot z = e$. Answer: $u =$ _____
   (e) $w \cdot u \cdot x = z$. Answer: $u =$ _____
   (f) $y \cdot u \cdot w = v$. Answer: $u =$ _____

43. Let $(G, \cdot)$ be any group. Solve each of the following equations for x, in terms of the group elements $a, b, c, a^{-1}, b^{-1}, c^{-1}$.

   (a) *Sample.* $a \cdot x \cdot b = c$.

   Solution. If x is a group element such that
   $a \cdot x \cdot b = c$, then $a^{-1} \cdot a \cdot x \cdot b = a^{-1} \cdot c$
   $$x \cdot b = a^{-1} \cdot c$$
   $$x \cdot b \cdot b^{-1} = a^{-1} \cdot c \cdot b^{-1}$$
   $$x = a^{-1} \cdot c \cdot b^{-1}$$

Groups 157

To check the solution: $a \cdot x \cdot b = a \cdot (a^{-1} \cdot c \cdot b^{-1}) \cdot b$
$= aa^{-1} \cdot c \cdot b^{-1} b$
$= e \cdot c \cdot e$
$= c$

$a^{-1}cb$

$b^{-1}ab^{-1}c^{-1}$

$b^{-1}aab^{-1}c$

$cbaacb^{-1}$

(b) $a \cdot x = cb$. Answer: $x = $ _____

(c) $x \cdot cba^{-1} = b^{-1}$. Answer: $x = $ _____

(d) $a^{-1}bxc^{-1}b = a$. Answer: $x = $ _____

(e) $a^{-1}b^{-1}c^{-1}xbc^{-1}a^{-1} = e$. Answer: $x = $ _____

44. Let $a, b \in Z$ and suppose $a \equiv b$ (m) where $m \in Z$ and $m > 1$. Prove that if $d \mid m$, then a, b have the same remainder when divided by d.

Proof. Since $a \equiv b$ (m), a, b, have the same remainder when divided by m. Therefore, there are integers q_1, q_2, r such that

$$a = mq_1 + r \text{ and } b = mq_2 + r$$

But $d \mid m$, so there is an integer k such that $m = dk$. Substituting $d \cdot k$ for m in the above equations gives

Eq. (1) $a = d(kq_1) + r$ and $b = d(kq_2) + r$

Now, if $r < d$, these equations show that a, b have the same remainder when divided by d. If $r \not< d$, then $r \geq d$. We may then write

Eq. (2) $r = d \cdot q_3 + r_1$

for some integers q_3 and r_1, where $0 \leq r_1 < d$. Use Eq. (2) and substitute for r in Eq. (1) to obtain

$a = d(kq_1) + dq_3 + r_1$ and $b = d(kq_2) + dq_3 + r_1$
$= d(kq_1 + q_3) + r_1$ $= d(kq_2 + q_3) + r_1$

In this case, r_1 is the remainder when a, b is divided by d. Hence, the proof is complete.

45. Let x, y, m be integers such that $m > 1$, $x \equiv y$ (m) and GCD $(x, m) = 1$. Show that GCD$(y, m) = 1$.

Solution. Suppose that d is a positive integer such that $d \mid y$ and $d \mid m$. Since $d \mid y$, we may conclude from the definition of "divides" that $y = d \cdot k$ for some integer k. This means that the remainder, when y is divided by d, is 0. Since $x \equiv y$ (m), x, y have the same remainder when divided by m and, by Problem 44 above, x, y have the same remainder when divided by d because $d \mid m$. Hence,

the remainder when x is divided by d is 0, i.e., $d|x$. Therefore, $d|m$ and $d|x$, which implies that $d = 1$ since $\text{GCD}(x, m) = 1$. This shows that if $d|y$ and $d|m$, then $d = 1$ so that $\text{GCD}(y, m) = 1$.

46. Let a, b, m be integers. Show that if $\text{GCD}(a, m) = 1$ and $\text{GCD}(b, m) = 1$, then $\text{GCD}(a \cdot b, m) = 1$.

Solution. Let d be a positive integer such that $d|m$ and $d|a \cdot b$. If $d \neq 1$, then there is a prime p such that $p|d$ by the Fundamental Theorem of Arithmetic. Then we have: $p|d$, $d|m$ and $d|a \cdot b$. From this it follows that $p|m$ and $p|a \cdot b$. But if $p|a \cdot b$, then $p|a$ or $p|b$. Hence, $p|m$ and $p|a$ or $p|m$ and $p|b$. Since we are given that $\text{GCD}(a, m) = 1$ and $\text{GCD}(b, m) = 1$ it is not possible for $p|m$ and $p|a$ or $p|m$ and $p|b$. Therefore, we have a contradiction and conclude that $d = 1$. This shows that if $d|m$ and $d|a \cdot b$, then $d = 1$. Thus $\text{GCD}(m, a \cdot b) = 1$.

47. Let m be an integer such that $m > 1$ and define $P_m = \{\bar{x} \mid \bar{x} \in Z_m^*, 0 \leq x < m$ and $\text{GCD}(x, m) = 1\}$. For example, if $m = 8$, $P_8 = \{\bar{1}, \bar{3}, \bar{5}, \bar{7}\}$. If we restrict the binary operation $\otimes$ on Z_m^* to P_m we obtain a binary operation on P_m. This is simply to say that if $\bar{x} \in P_m$ and $\bar{y} \in P_m$, then $\bar{x} \otimes \bar{y} \in P_m$. To illustrate, take $\bar{3} \in P_8$ and $\bar{5} \in P_8$. Then

$$\bar{3} \otimes \bar{5} = \overline{15} = \bar{7} \in P_8$$

We give below the table for $\otimes$ on P_8.

$\otimes$	$\bar{1}$	$\bar{3}$	$\bar{5}$	$\bar{7}$
$\bar{1}$	$\bar{1}$	$\bar{3}$	$\bar{5}$	$\bar{7}$
$\bar{3}$	$\bar{3}$	$\bar{1}$	$\bar{7}$	$\bar{5}$
$\bar{5}$	$\bar{5}$	$\bar{7}$	$\bar{1}$	$\bar{3}$
$\bar{7}$	$\bar{7}$	$\bar{5}$	$\bar{3}$	$\bar{1}$

It can be easily shown that $(P_8, \otimes)$ is a group. In fact, $(P_8, \otimes)$ is the so called Klein Four-group, named for the mathematician Felix Klein. We will now show that $(P_m, \otimes)$ is a group for all integers $m > 1$.

(a) $\otimes$ *is a binary operation on* P_m. We need to show that if $\bar{a} \in P_m$ and $\bar{b} \in P_m$, then $\bar{a} \otimes \bar{b} \in P_m$. If $\bar{a} \in P_m$ and $\bar{b} \in P_m$, we have, by definition of P_m, that $\text{GCD}(a, m) = 1$, $\text{GCD}(b, m) = 1$, $0 < a < m$ and $0 < b < m$. Then, by Problem 45 above, $\text{GCD}(a \cdot b, m) = 1$. It is possible that $a \cdot b \geq m$. But there is an integer r such that $0 \leq r < m$ and $a \cdot b \equiv r \ (m)$. Then $\overline{a \cdot b} = \bar{r}$ and, by Problem 44 above, $\text{GCD}(r, m) = 1$ since $\text{GCD}(a \cdot b, m) = 1$. Hence, $\bar{r} \in P_m$ and

$$\bar{a} \otimes \bar{b} = \overline{a \cdot b} = \bar{r} \in P_m$$

(b) $\otimes$ is associative on P_m. This has already been shown.
(c) $\bar{1}$ is the identity element of P_m.
(d) Each element $\bar{g} \in P_m$ has an inverse.

Since $GCD(g, m) = 1$, there are integers x, y such that $gx + my = 1$. Then if d is a positive integer such that $d \mid x$ and $d \mid m$, $d \mid 1$. Hence, $GCD(x, m) = 1$. Also from $gx + my = 1$ it follows that $gx \equiv 1(m)$. It may happen that $x \geq m$ or $x < 0$. But it is possible to find an integer r such that $0 < r < m$ and $x \equiv r(m)$. Since $GCD(x, m) = 1$ and $x \equiv r\ (m)$, $GCD(r, m) = 1$ by Problem 44. Then $gx \equiv gr(m)$ and since $gx \equiv 1(m)$, $gr \equiv 1(m)$. Therefore,

$$\bar{g} \otimes \bar{r} = \overline{gr} = \bar{1}$$

so that $\bar{r} \in P_m$ and is the inverse of $\bar{g}$.

48. Let n be a natural number and let $a_1, a_2, a_3, \ldots, a_n$ be elements of a group $(G, \cdot)$. Prove by induction that

$$(a_1 \cdot a_2 \cdot a_3 \cdot \ldots \cdot a_n)^{-1} = a_n^{-1} a_{n-1}^{-1} a \ldots a_3^{-1} a_2^{-1} a_1^{-1}$$

49. Complete the proof of Theorem 22 by showing that the solution ba^{-1} for the equation $ya = b$ is unique.

50. Complete the proof of Theorem 23 by showing that L_a is a 1–1 mapping of G onto G for each $a \in G$.

4.4 Cyclic Groups and Subgroups

Exponent notation may be introduced into groups in a natural way so that some of the usual laws of exponents hold. If a is an element of a group $(G, \cdot)$ and n is a positive integer, we define a^n to mean the group element $a \cdot a \cdot a \ldots a$, i.e.,

$$a^n = \underbrace{a \cdot a \cdot a \ldots a}_{n \text{ factors of } a}$$

a^0 is defined to be the identity element of the group, $a^0 = e$, and a^{-n} is defined to be $(a^n)^{-1}$ if $n > 0$. For example

$$a^3 = a \cdot a \cdot a \text{ and } a^{-3} = (a^3)^{-1} = (a \cdot a \cdot a)^{-1} = a^{-1} \cdot a^{-1} \cdot a^{-1}$$

The following laws of exponents may be proved from these definitions.
(1) $g^n \cdot g^m = g^{n+m}$ for all $g \in G$ and $n, m \in Z$.

(2) $g^n \cdot g^{-m} = g^{n-m}$ for all $g \in G$ and $n, m \in Z$.
(3) $(g^n)^m = g^{n \cdot m}$ for all $g \in G$ and $n, m \in Z$.

If $g \in G$, where $(G, \cdot)$ is a group, the elements of G

$$\ldots, g^{-3}, g^{-2}, g^{-1}, g^0, g^1, g^2, g^3, \ldots$$

are called the *powers* of g. Note that every power of the identity element is the identity. Concerning the powers of a group element g, two possibilities arise: either all the powers of g are distinct, i.e., $g^n \neq g^m$ for all integers n, m such that $n \neq m$, or $g^n = g^m$ for some integers n, m such that $n \neq m$. Suppose that $n > m$ and $g^n = g^m$. Then $g^n \cdot g^{-m} = g^m \cdot g^{-m}$ and

$$g^{n-m} = e$$

Since $n > m$, $n - m > 0$ and we may say that *some positive power of g is the identity element in this case*. As an example of this situation consider $\bar{2} \in Z_{11}^*$. The powers of $\bar{2}$ are

$$\bar{2}^0 = \bar{1}$$
$$\bar{2}^1 = \bar{2}$$
$$\bar{2}^2 = \bar{2} \otimes \bar{2} = \bar{4}$$
$$\bar{2}^3 = \bar{2}^2 \otimes \bar{2} = \bar{4} \otimes \bar{2} = \bar{8}$$
$$\bar{2}^4 = \bar{2}^3 \otimes \bar{2} = \bar{8} \otimes \bar{2} = \bar{5}$$
$$\bar{2}^5 = \bar{2}^4 \otimes \bar{2} = \bar{5} \otimes \bar{2} = \overline{10}$$
$$\bar{2}^6 = \bar{2}^5 \otimes \bar{2} = \overline{10} \otimes \bar{2} = \bar{9}$$
$$\bar{2}^7 = \bar{2}^6 \otimes \bar{2} = \bar{9} \otimes \bar{2} = \bar{7}$$
$$\bar{2}^8 = \bar{2}^7 \otimes \bar{2} = \bar{7} \otimes \bar{2} = \bar{3}$$
$$\bar{2}^9 = \bar{2}^8 \otimes \bar{2} = \bar{3} \otimes \bar{2} = \bar{6}$$
$$\bar{2}^{10} = \bar{2}^9 \otimes \bar{2} = \bar{6} \otimes \bar{2} = \bar{1}$$

Here, the 10th power of $\bar{2}$ is the group identity element.

In the group $(G, \cdot)$, where G is the set of positive rational numbers and $\cdot$ is ordinary multiplication, no positive power of the element $\frac{1}{2}$, for example, is the group identity 1. If we had $(\frac{1}{2})^n = 1$ for some natural number n, it would follow that $\frac{1}{2^n} = 1$ and $2^n = 1$. But $2^n \neq 1$ for all natural numbers n.

DEFINITION. Let g be an element of a group $(G, \cdot)$. If some positive power of g is the identity element of G, then the smallest positive integer n such that $g^n = e$ is called the *order* of the element g. If no positive power of g is the identity element of G, then g is said to have *infinite order*. An element is said to have *finite order* if it does not have infinite order.

From the example above, we see that for $\bar{2} \in Z_{11}^*$, 10 is the smallest positive integer such that $\bar{2}^{10} = \bar{1}$. Hence, the order of $\bar{2}$ is 10. For $\bar{5} \in Z_{11}^*$ we have

$$\bar{5}^1 = \bar{5}$$
$$\bar{5}^2 = \bar{5} \otimes \bar{5} = \bar{3}$$
$$\bar{5}^3 = \bar{5}^2 \otimes \bar{5} = \bar{3} \otimes \bar{5} = \bar{4}$$
$$\bar{5}^4 = \bar{5}^3 \otimes \bar{5} = \bar{4} \otimes \bar{5} = \bar{9}$$
$$\bar{5}^5 = \bar{5}^4 \otimes \bar{5} = \bar{9} \otimes \bar{5} = \bar{1}$$

Therefore, the order of $\bar{5}$ is 5.

Theorem 24. Let $(G, \cdot)$ be a group and let $g \in G$ such that g has finite order n. Then if $g^m = e$, $n \mid m$.

Proof. Apply the division algorithm and write $m = nq + r$ for some integers q, r such that $0 \leq r < n$. Then

$$\begin{aligned}
e = g^m &= g^{nq+r} \\
&= g^{nq} \cdot g^r \\
&= (g^n)^q \cdot g^r \\
&= e^q \cdot g^r \qquad \text{since } g^n = e \\
&= e \cdot g^r \\
&= g^r
\end{aligned}$$

Hence, $g^r = e$ and $r < n$. Since n is the smallest positive integer such that $g^n = e$, we conclude that $r = 0$. Therefore, $m = nq + r = nq + 0 = nq$ which implies that $n \mid m$.

Consider the group table below of $(Z_6, \oplus)$. Certain blocks have been shaded in the table to focus attention upon the subset $H = \{\bar{0}, \bar{2}, \bar{4}\}$ of Z_6. We observe from the table that if $\bar{x}, \bar{y} \in H$, then $\bar{x} \oplus \bar{y} \in H$ which tells us that $\oplus$ is a binary operation

$\oplus$	$\bar{0}$	$\bar{1}$	$\bar{2}$	$\bar{3}$	$\bar{4}$	$\bar{5}$
$\bar{0}$	$\bar{0}$	$\bar{1}$	$\bar{2}$	$\bar{3}$	$\bar{4}$	$\bar{5}$
$\bar{1}$	$\bar{1}$	$\bar{2}$	$\bar{3}$	$\bar{4}$	$\bar{5}$	$\bar{0}$
$\bar{2}$	$\bar{2}$	$\bar{3}$	$\bar{4}$	$\bar{5}$	$\bar{0}$	$\bar{1}$
$\bar{3}$	$\bar{3}$	$\bar{4}$	$\bar{5}$	$\bar{0}$	$\bar{1}$	$\bar{2}$
$\bar{4}$	$\bar{4}$	$\bar{5}$	$\bar{0}$	$\bar{1}$	$\bar{2}$	$\bar{3}$
$\bar{5}$	$\bar{5}$	$\bar{0}$	$\bar{1}$	$\bar{2}$	$\bar{3}$	$\bar{4}$

on H. The identity element $\bar{0} \in H$ and each element of H has an inverse in H. It is obvious that $\oplus$ is associative on H since $\oplus$ is associative on all Z_6. Hence, we conclude that $(H, \oplus)$ is itself a group. We describe this situation by saying that $(H, \oplus)$ is a *subgroup* of $(Z_6, \oplus)$.

DEFINITION. Let $(G, \cdot)$ be a group and let $H \subseteq G$. Then $(H, \cdot)$ is called a *subgroup* of $(G, \cdot)$ if $(H, \cdot)$ is a group.

According to the definition, $(H, \cdot)$ is a subgroup of $(G, \cdot)$ if $(H, \cdot)$ is a group. This implies that $\cdot$ is a binary operation on H, i.e., $a \cdot b \in H$ if $a, b \in H$. Also $H \neq \varnothing$ since H must have an identity element and each element $a \in H$ must have its inverse in H. Hence, choose any element $a \in H$; then $a^{-1} \in H$ so that $a \cdot a^{-1} \in H$, i.e., $e \in H$. Since the associative law holds for all of G, it holds automatically in H. We conclude that $(H, \cdot)$ is a subgroup of $(G, \cdot)$, where $H \subseteq G$, if and only if
(1) $a \cdot b \in H$ for all $a, b \in H$;
(2) the identity e of G is in H;
(3) for each $g \in H$, $g^{-1} \in H$.

Example 4-8: Consider the group $(Z, +)$. The subset of Z consisting of all multiples of 2, i.e., $\{2n \mid n \in Z\}$ is a subgroup of $(Z, +)$. (1) $2n + 2m = 2(n + m)$ shows that $+$ is a binary operation on the set of all multiples of 2. (2) The identity of $(Z, +)$ is 0. But 0 is a multiple of 2: $0 = 2 \cdot 0$. (3) For each $2n$, the inverse is $2(-n)$.

Example 4-9: For any group $(G, \cdot)$, $(G, \cdot)$ is a subgroup of $(G, \cdot)$. Also $(\{e\}, \cdot)$ is a subgroup of $(G, \cdot)$, where e is the identity element of $(G, \cdot)$.

Example 4-10: Define $H = \{a/2^n \mid a \in Z, n \in Z, a > 0\}$. This is a subset of the positive rational numbers. We have seen that the positive rational numbers are a group with respect to ordinary multiplication. We now check to see if $(H, \cdot)$ is a subgroup.
(1) If $a/2^n \in H$ and $b/2^m \in H$, then $(a/2^n) \cdot (b/2^m) = (ab)/(2^{n+m}) \in H$ which shows that $\cdot$ is a binary operation on H.
(2) The identity element of the positive rational numbers with respect to multiplication is 1. Since $1 = \frac{1}{2^0}$, $1 \in H$.
(3) By definition of H, $3/2 \in H$ and the inverse of $3/2$ is $2/3$. But $2/3 \notin H$. Therefore, $(H, \cdot)$ is not a subgroup of the positive rational numbers with respect to multiplication.

Example 4-11: Let $(G, \cdot)$ be a group and let $g \in G$. Define $H = \{g^n \mid n \in Z\}$. Then $(H, \cdot)$ is a subgroup of $(G, \cdot)$.
(1) For $g^n, g^m \in H$, $g^n \cdot g^m = g^{n+m} \in H$.
(2) Since $e = g^0$ by definition, $e \in H$.
(3) For $g^n \in H$, $g^{-n} \in H$ and $g^n \cdot g^{-n} = g^0 = e$.

Theorem 25. Let $(G, \cdot)$ be a group and let $(H, \cdot)$ and $(K, \cdot)$ be subgroups of $(G, \cdot)$. Then $(H \cap K, \cdot)$ is a subgroup of $(G, \cdot)$.

Proof. (1) If $a \in H \cap K$ and $b \in H \cap K$, then $a \in H$, $a \in K$, $b \in H$ and $b \in K$

by definition of intersection. Since $a, b \in H$, $a \cdot b \in H$, because $(H, \cdot)$ is a subgroup. Also since $(K, \cdot)$ is a subgroup and $a, b \in K$, $a \cdot b \in K$. Therefore, since $a \cdot b \in H$ and $a \cdot b \in K$, $a \cdot b \in H \cap K$.

(2) Both $(H, \cdot)$ and $(K, \cdot)$ are subgroups and hence $e \in H$ and $e \in K$. Thus $e \in H \cap K$.

(3) If $g \in H \cap K$, then $g \in H$ and $g \in K$. Then, since $(H, \cdot)$ and $(K, \cdot)$ are subgroups, $g^{-1} \in H$ and $g^{-1} \in K$. It then follows that $g^{-1} \in H \cap K$. Therefore, $(H \cap K, \cdot)$ is a subgroup of $(G, \cdot)$.

Corollary. If M is any set of subgroups of a group $(G, \cdot)$, then the intersection of the members of M is a subgroup of $(G, \cdot)$.

If we change $\cap$ to $\cup$ in Theorem 25, the resulting statement is not true. However, there is a "smallest" subgroup of $(G, \cdot)$ which contains $H \cup K$, where $(H, \cdot)$ and $(K, \cdot)$ are any two subgroups of $(G, \cdot)$.

Theorem 26. Let $(G, \cdot)$ be a group and let S be any subset of G. Then there is a subgroup $(T, \cdot)$ of $(G, \cdot)$ such that
(a) $S \subseteq T$; and
(b) if $(K, \cdot)$ is any subgroup of $(G, \cdot)$ such that $S \subseteq K$, then $T \subseteq K$.

Proof. There is at least one subgroup of $(G, \cdot)$ which contains the set S, namely $(G, \cdot)$ itself. Therefore,

$$M = \{(H, \cdot) \mid (H, \cdot) \text{ is a subgroup of } (G, \cdot) \text{ and } S \subseteq H\}$$

is not the null set. Now we let T be the intersection of all the H such that $(H, \cdot) \in M$. We know that if A, B are any two sets such that $S \subseteq A$ and $S \subseteq B$, then $S \subseteq (A \cap B)$. Therefore, since for each $(H, \cdot) \in M$, $S \subseteq H$ and T is the intersection all these H, $S \subseteq T$. Also, $(T, \cdot)$ is a subgroup of $(G, \cdot)$ by the corollary to Theorem 25. This proves (a). Now suppose that $(K, \cdot)$ is a subgroup of $(G, \cdot)$ such that $S \subseteq K$. Then by definition of M, $(K, \cdot) \in M$ and by definition of T, $T \subseteq K$ (by the result that $(A \cap B) \subseteq A$ and $(A \cap B) \subseteq B$). This proves (b).

The subgroup $(T, \cdot)$ of Theorem 26 is called the *subgroup generated by S* and is denoted by $([S], \cdot)$ or just $[S]$. If $S = \{g\}$ for $g \in G$, then $[\{g\}]$ is called the *cyclic subgroup generated by g*. This notation $[\{g\}]$ is usually shortened to simply $[g]$. It was shown above that if $H = \{g^n \mid n \in Z\}$, then $(H, \cdot)$ is a subgroup of $(G, \cdot)$. Since $g^1 = g$, $g \in H$, and by (b) of Theorem 26 it follows that $[g] \subseteq H$. But $[g]$ must contain all integral powers of g: since $g \in [g]$ and $[g]$ is a group, $g \cdot g = g^2 \in [g]$, $g^2 \cdot g = g^3 \in [g]$, etc., and also $g^{-1} \in [g]$, $g^{-2} = (g^{-1})^2 = g^{-1} \cdot g^{-1} \in [g]$, $g^{-3} = g^{-2} \cdot g^{-1} \in [g]$, etc. This shows that $H \subseteq [g]$ and hence, that $[g] = H$. Therefore, the *cyclic subgroup generated by g, $[g]$, consists of all integral powers of g*. If $(G, \cdot)$ is a group and there is an element $g \in G$ such that $[g] = G$, then $(G, \cdot)$ is called a *cyclic group*.

Theorem 27. Let $(G, \cdot)$ be a cyclic group. Then every subgroup $(H, \cdot)$ of $(G, \cdot)$ is cyclic.

Proof. Suppose that $G = [g]$ for $g \in G$. If $H = \{e\}$, then H is the cyclic subgroup $[e]$. If $H \neq \{e\}$, then there is some $h \in H$ and $h \neq e$. Since $h \in G = [g]$, $h = g^n$ for some $n \in Z$ with $n \neq 0$. H is a group so that $h^{-1} = g^{-n} \in H$. Then either n or $-n$ is positive since $n \neq 0$. Let m be the smallest positive integer such that $g^m \in H$. Let $j = g^m$. Then $[j]$ is a subgroup of H, $[j] \subseteq H$. If $h = g^n$ is any element of H, apply the Division Algorithm to m, n:

$$n = mq + r, \quad 0 \leq r < m$$

for some integers q, r. Then

$$\begin{aligned} h = g^n &= g^{mq+r} \\ &= g^{mq} \cdot g^r \\ &= (g^m)^q \cdot g^r \\ &= j^q \cdot g^r \end{aligned}$$

Hence, $g^r = j^{-q} \cdot h$. Since $j^{-q} \in H$ and $h \in H$, then $g^r \in H$. But $0 \leq r < m$ and m is the smallest positive integer such that $g^m \in H$. We conclude that $r = 0$. Therefore, $n = mq$ and

$$\begin{aligned} h = g^n &\\ &= g^{mq} = (g^m)^q \\ &= j^q \in [j] \end{aligned}$$

This shows that if $h \in H$, then $h \in [j]$. Thus it follows that $H = [j]$, showing that H is a cyclic subgroup of $(G, \cdot)$.

Example 4-12: We compute all the subgroups of $(Z_6, \oplus)$. We refer here to the group table given on p. 161. First, the most obvious subgroups are $(Z_6, \oplus)$ and $(\{\bar{0}\}, \oplus)$. Since no confusion will arise, we will denote subgroups by the sets only. For example, Z_6 and $\{\bar{0}\}$ stand for $(Z_6, \oplus)$ and $(\{\bar{0}\}, \oplus)$ respectively. Next, we compute the cyclic subgroups $[\bar{0}]$, $[\bar{1}]$, $[\bar{2}]$, $[\bar{3}]$, $[\bar{4}]$, $[\bar{5}]$. Remember that these consist of integral powers of the generating element.

$$\begin{aligned} [\bar{0}] &= \{\bar{0}\}; \\ [\bar{1}] &= \{\bar{0}, \bar{1}, \bar{2}, \bar{3}, \bar{4}, \bar{5}\} = Z_6 \\ [\bar{2}] &= \{\bar{0}, \bar{2}, \bar{4}\} \\ [\bar{3}] &= \{\bar{0}, \bar{3}\} \\ [\bar{4}] &= \{\bar{0}, \bar{4}, \bar{2}\} = [\bar{2}] \\ [\bar{5}] &= \{\bar{0}, \bar{5}, \bar{4}, \bar{3}, \bar{2}, \bar{1}\} = Z_6 \end{aligned}$$

Observe that Z_6 is a cyclic group and that by Theorem 27 the above subgroups are the only ones.

Exercises

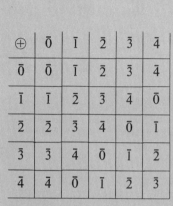

{$\bar{0}$}
{$\bar{1}, \bar{2}, \bar{3}, \bar{4}, \bar{0}$}
{$\bar{2}, \bar{4}, \bar{1}, \bar{3}, \bar{0}$}
{$\bar{3}, \bar{1}, \bar{4}, \bar{2}, \bar{0}$}
{$\bar{4}, \bar{3}, \bar{2}, \bar{1}, \bar{0}$}
Yes

$\bar{1}, \bar{2}, \bar{3}, \bar{4}$

No
Cyclic groups have only cyclic subgroups.
1
5
5

51. Construct the group table for $(Z_5, \oplus)$ below and compute the cyclic subgroups.

$\oplus$	$\bar{0}$	$\bar{1}$	$\bar{2}$	$\bar{3}$	$\bar{4}$
$\bar{0}$					
$\bar{1}$					
$\bar{2}$					
$\bar{3}$					
$\bar{4}$					

$[\bar{0}] =$ _____

$[\bar{1}] =$ _____

$[\bar{2}] =$ _____

$[\bar{3}] =$ _____

$[\bar{4}] =$ _____

(a) Is $(Z_5, \oplus)$ a cyclic group? _____

(b) What are the elements $\bar{a} \in Z_5$ such that $[\bar{a}] = Z_5$? _____

(c) Does $(Z_5, \oplus)$ have any subgroups other than those listed above? _____
Why? _____

(d) What is the order of $\bar{0}$? _____

(e) What is the order of $\bar{1}$? _____

(f) What are the orders of $\bar{2}, \bar{3}, \bar{4}$? _____

52. Construct the group table for $(P_8, \otimes)$—see Problem 47, p. 158—below and compute the cyclic subgroups.

$\otimes$	$\bar{1}$	$\bar{3}$	$\bar{5}$	$\bar{7}$
$\bar{1}$				
$\bar{3}$				
$\bar{5}$				
$\bar{7}$				

[$\bar{1}$] = _____

[$\bar{5}$] = _____

[$\bar{3}$] = _____

[$\bar{7}$] = _____

(a) Is $(P_8, \otimes)$ a cyclic group? _____

Why? _____

(b) What is the order of $\bar{1}$? _____

(c) What are the orders of $\bar{3}, \bar{5}, \bar{7}$? _____

(d) Does [$\bar{3}, \bar{5}$] contain $\bar{3}$ and $\bar{5}$? _____

(e) Does [$\bar{3}, \bar{5}$] contain $\bar{3} \otimes \bar{5}$? _____

(f) [$\bar{3}, \bar{5}$] = _____

(g) [$\bar{3}, \bar{7}$] = _____

(h) [$\bar{5}, \bar{7}$] = _____

$\otimes$	$\bar{1}$	$\bar{3}$	$\bar{5}$	$\bar{7}$
$\bar{1}$	$\bar{1}$	$\bar{3}$	$\bar{5}$	$\bar{7}$
$\bar{3}$	$\bar{3}$	$\bar{1}$	$\bar{7}$	$\bar{5}$
$\bar{5}$	$\bar{5}$	$\bar{7}$	$\bar{1}$	$\bar{3}$
$\bar{7}$	$\bar{7}$	$\bar{5}$	$\bar{3}$	$\bar{1}$

{$\bar{1}$}

{$\bar{5}, \bar{1}$}

{$\bar{3}, \bar{1}$}

{$\bar{7}, \bar{1}$}

No

None of the sets [$\bar{1}$], [$\bar{3}$], [$\bar{5}$], [$\bar{7}$], is P_8.

1

2

Yes

Yes

{$\bar{3}, \bar{1}, \bar{5}, \bar{7}$}

{$\bar{3}, \bar{1}, \bar{7}, \bar{5}$}

{$\bar{5}, \bar{1}, \bar{7}, \bar{3}$}

53. Construct the group table below for $(Z_{12}, \oplus)$.

$\oplus$	$\bar{0}$	$\bar{1}$	$\bar{2}$	$\bar{3}$	$\bar{4}$	$\bar{5}$	$\bar{6}$	$\bar{7}$	$\bar{8}$	$\bar{9}$	$\overline{10}$	$\overline{11}$
$\bar{0}$												
$\bar{1}$												
$\bar{2}$												
$\bar{3}$												
$\bar{4}$												
$\bar{5}$												
$\bar{6}$												
$\bar{7}$												
$\bar{8}$												
$\bar{9}$												
$\overline{10}$												
$\overline{11}$												

Groups

$\oplus$	$\bar{0}$	$\bar{1}$	$\bar{2}$	$\bar{3}$	$\bar{4}$	$\bar{5}$	$\bar{6}$	$\bar{7}$	$\bar{8}$	$\bar{9}$	$\overline{10}$	$\overline{11}$
$\bar{0}$	$\bar{0}$	$\bar{1}$	$\bar{2}$	$\bar{3}$	$\bar{4}$	$\bar{5}$	$\bar{6}$	$\bar{7}$	$\bar{8}$	$\bar{9}$	$\overline{10}$	$\overline{11}$
$\bar{1}$	$\bar{1}$	$\bar{2}$	$\bar{3}$	$\bar{4}$	$\bar{5}$	$\bar{6}$	$\bar{7}$	$\bar{8}$	$\bar{9}$	$\overline{10}$	$\overline{11}$	$\bar{0}$
$\bar{2}$	$\bar{2}$	$\bar{3}$	$\bar{4}$	$\bar{5}$	$\bar{6}$	$\bar{7}$	$\bar{8}$	$\bar{9}$	$\overline{10}$	$\overline{11}$	$\bar{0}$	$\bar{1}$
$\bar{3}$	$\bar{3}$	$\bar{4}$	$\bar{5}$	$\bar{6}$	$\bar{7}$	$\bar{8}$	$\bar{9}$	$\overline{10}$	$\overline{11}$	$\bar{0}$	$\bar{1}$	$\bar{2}$
$\bar{4}$	$\bar{4}$	$\bar{5}$	$\bar{6}$	$\bar{7}$	$\bar{8}$	$\bar{9}$	$\overline{10}$	$\overline{11}$	$\bar{0}$	$\bar{1}$	$\bar{2}$	$\bar{3}$
$\bar{5}$	$\bar{5}$	$\bar{6}$	$\bar{7}$	$\bar{8}$	$\bar{9}$	$\overline{10}$	$\overline{11}$	$\bar{0}$	$\bar{1}$	$\bar{2}$	$\bar{3}$	$\bar{4}$
$\bar{6}$	$\bar{6}$	$\bar{7}$	$\bar{8}$	$\bar{9}$	$\overline{10}$	$\overline{11}$	$\bar{0}$	$\bar{1}$	$\bar{2}$	$\bar{3}$	$\bar{4}$	$\bar{5}$
$\bar{7}$	$\bar{7}$	$\bar{8}$	$\bar{9}$	$\overline{10}$	$\overline{11}$	$\bar{0}$	$\bar{1}$	$\bar{2}$	$\bar{3}$	$\bar{4}$	$\bar{5}$	$\bar{6}$
$\bar{8}$	$\bar{8}$	$\bar{9}$	$\overline{10}$	$\overline{11}$	$\bar{0}$	$\bar{1}$	$\bar{2}$	$\bar{3}$	$\bar{4}$	$\bar{5}$	$\bar{6}$	$\bar{7}$
$\bar{9}$	$\bar{9}$	$\overline{10}$	$\overline{11}$	$\bar{0}$	$\bar{1}$	$\bar{2}$	$\bar{3}$	$\bar{4}$	$\bar{5}$	$\bar{6}$	$\bar{7}$	$\bar{8}$
$\overline{10}$	$\overline{10}$	$\overline{11}$	$\bar{0}$	$\bar{1}$	$\bar{2}$	$\bar{3}$	$\bar{4}$	$\bar{5}$	$\bar{6}$	$\bar{7}$	$\bar{8}$	$\bar{9}$
$\overline{11}$	$\overline{11}$	$\bar{0}$	$\bar{1}$	$\bar{2}$	$\bar{3}$	$\bar{4}$	$\bar{5}$	$\bar{6}$	$\bar{7}$	$\bar{8}$	$\bar{9}$	$\overline{10}$

$\{\bar{0}\}$

$\{\bar{1}, \bar{2}, \bar{3}, \bar{4}, \bar{5}, \bar{6}, \bar{7}, \bar{8}, \bar{9}, \overline{10}, \overline{11}, \bar{0}\}$

$\{\bar{2}, \bar{4}, \bar{6}, \bar{8}, \overline{10}, \bar{0}\}$

$\{\bar{3}, \bar{6}, \bar{9}, \bar{0}\}$

$\{\bar{4}, \bar{8}, \bar{0}\}$

$\{\bar{5}, \overline{10}, \bar{3}, \bar{8}, \bar{1}, \bar{6}, \overline{11}, \bar{4}, \bar{9}, \bar{2}, \bar{7}, \bar{0}\}$

$\{\bar{6}, \bar{0}\}$

$\{\bar{7}, \bar{2}, \bar{9}, \bar{4}, \overline{11}, \bar{6}, \bar{1}, \bar{8}, \bar{3}, \overline{10}, \bar{5}, \bar{0}\}$

$\{\bar{8}, \bar{4}, \bar{0}\}$

$\{\bar{9}, \bar{6}, \bar{3}, \bar{0}\}$

$\{\overline{10}, \bar{8}, \bar{6}, \bar{4}, \bar{2}, \bar{0}\}$

$\{\overline{11}, \overline{10}, \bar{9}, \bar{8}, \bar{7}, \bar{6}, \bar{5}, \bar{4}, \bar{3}, \bar{2}, \bar{1}, \bar{0}\}$

$[\bar{0}] =$ _____

$[\bar{1}] =$ _____

$[\bar{2}] =$ _____

$[\bar{3}] =$ _____

$[\bar{4}] =$ _____

$[\bar{5}] =$ _____

$[\bar{6}] =$ _____

$[\bar{7}] =$ _____

$[\bar{8}] =$ _____

$[\bar{9}] =$ _____

$[\overline{10}] =$ _____

$[\overline{11}] =$ _____

Groups   169

Yes	(a) Is $(Z_{12}, \oplus)$ a cyclic group?
$\bar{1}, \bar{5}, \bar{7}, \overline{11}$	(b) What elements generate Z_{12}?
1	(c) The order of $\bar{0}$ is
12	(d) The order of $\bar{1}$ is
6	(e) The order of $\bar{2}$ is
4	(f) The order of $\bar{3}$ is
3	(g) The order of $\bar{4}$ is
12	(h) The order of $\bar{5}$ is
2	(i) The order of $\bar{6}$ is
12	(j) The order of $\bar{7}$ is
3	(k) The order of $\bar{8}$ is
4	(l) The order of $\bar{9}$ is
6	(m) The order of $\overline{10}$ is
12	(n) The order of $\overline{11}$ is

(o) Does $(Z_{12}, \oplus)$ have any subgroups other than those listed above?

No

Why?

Cyclic groups have only cyclic subgroups.

(p) Compute the following intersections:

$\{\bar{0}, \bar{6}\} = [\bar{6}]$
  (i) $[\bar{2}] \cap [\bar{3}] = $ _____ = _____

$\{\bar{4}, \bar{8}, \bar{0}\} = [\bar{4}]$
  (ii) $[\bar{2}] \cap [\bar{4}] = $ _____ = _____

$\{\bar{0}\} = [\bar{0}]$
  (iii) $[\bar{3}] \cap [\bar{4}] = $ _____ = _____

$\{\bar{0}, \bar{6}\} = [\bar{6}]$
  (iv) $[\bar{3}] \cap [\bar{6}] = $ _____ = _____

$\{\bar{0}, \bar{6}\} = [\bar{6}]$
  (i) $[\bar{3}] \cap [\overline{10}] = $ _____ = _____

$\{\bar{0}, \bar{4}, \bar{8}\} = [\bar{4}]$
  (vi) $[\bar{4}] \cap [\overline{10}] = $ _____ = _____

54. In $(Z_{19}^*, \otimes)$ compute $\bar{5}^3, \overline{10}^2, \overline{15}^3$.

(a) $\bar{5}^3 = \bar{5} \otimes \bar{5} \otimes \bar{5} = \overline{25} \otimes \bar{5} = \bar{6} \otimes \bar{5} = \overline{11}$.

$\overline{10} \otimes \overline{10}, \overline{100}, \overline{5}$

$\overline{15} \otimes \overline{15} \otimes \overline{15} = \overline{225} \otimes \overline{15}$

$\overline{16} \otimes \overline{15} = \overline{12}$

$\overline{14} \oplus \overline{14} \oplus \overline{14} = \overline{42} = \overline{12}$

$\overline{21} \oplus \overline{21} \oplus \overline{21} \oplus \overline{21} \oplus \overline{21} = \overline{105}$
$= \overline{15}$

7

11

6

3

36

H is a subgroup of G.

H is not a subgroup because the inverses of elements of H are not in H.

H is not a subgroup because $+$ is not a binary operation on H.

(b) $\overline{10^2} = $ _____ = _____ = _____

(c) $\overline{15^3} = $ _____ = _____
= _____ = _____

55. In $(Z_{30}, \oplus)$ compute $\overline{8}^6, \overline{14}^3, \overline{21}^5$.

(a) $\overline{8}^6 = \overline{8} \oplus \overline{8} \oplus \overline{8} \oplus \overline{8} \oplus \overline{8} \oplus \overline{8} = \overline{48} = \overline{18}$.

(b) $\overline{14}^3 = $ _____ = _____ = _____

(c) $\overline{21}^5 = $ _____ = _____ = _____

_____

56. (a) In $(Z_7^*, \otimes)$ compute the order of $\overline{5}$. _____

(b) In $(Z_{11}^*, \otimes)$ compute the order of $\overline{2}$. _____

(c) In $(Z_{36}, \oplus)$ compute the order of $\overline{6}$. _____

(d) In $(Z_{36}, \oplus)$ compute the order of $\overline{12}$. _____

(e) In $(Z_{36}, \oplus)$ compute the order of $\overline{5}$. _____

57. Decide whether each H below is a subgroup of the given G. If you decide that it is not a subgroup, give a reason. (a) G is the set of integers with binary operation of addition and H is the set of all integers which are multiples of 3. _____

(b) G is the set of integers with binary operation of addition and H is the set of all nonnegative integers. _____

_____

_____

(c) G is the set of integers with binary operation of addition and H is the set of odd integers. _____

_____

Groups   171

H is a subgroup.

H is not a subgroup of *G* because the inverse of elements of *H* are not in *H*.

H is a subgroup of *G*.

(d) *G* is the set of all rational numbers with binary operation addition and *H* is the subset of all integers. _____

(e) *G* is the set of all positive rational numbers with binary operation multiplication and *H* is the subset of natural numbers. _____
_____
_____

(f) *G* is the set of all real numbers with binary operation addition and *H* is the set of rational numbers. _____

58. Show that for every integer $m > 1$, $(Z_m, \oplus)$ is a cyclic group.

59. Let $(G, \cdot)$ be a cyclic group with generator g, i.e., $G = [g]$. Prove that if g has order n, for some integer $n > 1$, and if $m \in Z$ such $\text{GCD}(m, n) = 1$, then $G = [g^m]$.

60. Let $(G, \cdot)$ be a group and let $g \in G$. Define $H = \{x \mid x \cdot g = g \cdot x, x \in G\}$. Show that H is a subgroup of G.

61. Let $(G, \cdot)$ be a group and define $H = \{y \mid y \cdot x = x \cdot y \text{ for all } x \in G \text{ and } y \in G\}$. Show that H is a subgroup of G.

62. Prove the laws of exponents given on pp. 159–160.

4.5 Permutation Groups

We have already studied 1–1 mappings of a set X onto itself. These mappings are called *permutations* on the set X when X is finite. For example, when $X = \{1, 2, 3\}$, we

have the six permutations on X listed below:

$$I = \begin{pmatrix} 1 & 2 & 3 \\ 1 & 2 & 3 \end{pmatrix}, A = \begin{pmatrix} 1 & 2 & 3 \\ 1 & 3 & 2 \end{pmatrix}, B = \begin{pmatrix} 1 & 2 & 3 \\ 3 & 2 & 1 \end{pmatrix}$$

$$C = \begin{pmatrix} 1 & 2 & 3 \\ 2 & 1 & 3 \end{pmatrix}, D = \begin{pmatrix} 1 & 2 & 3 \\ 2 & 3 & 1 \end{pmatrix}, E = \begin{pmatrix} 1 & 2 & 3 \\ 3 & 1 & 2 \end{pmatrix}$$

Permutations on a set X furnish an extremely important example of groups.

Theorem 28. Let X be a non-empty set and let M be the set of all 1–1 mappings of X onto X. Then $(M, \cdot)$ is a group, where $\cdot$ is the product of mappings.

Proof. (a) Let $T \in M$ and $S \in M$. We must show that $T \cdot S \in M$ or that $T \cdot S$ is a 1–1 mapping of X onto X. We need to show that $T \cdot S$ is a mapping, that it is 1–1 and that it is onto. To show that it is a mapping, it must be shown that if $x = y$ and $T \cdot S: x \longrightarrow z$ and $T \cdot S: y \longrightarrow w$, then $z = w$. Suppose that

$$T: x \longrightarrow m \qquad S: m \longrightarrow z$$
$$T: y \longrightarrow n \qquad S: n \longrightarrow w$$

Now if $x = y$, we must have $m = n$ since T is a mapping and $z = w$ since S is a mapping. Then by definition of $T \cdot S$:

$$T \cdot S: x \underset{T}{\longrightarrow} m \underset{S}{\longrightarrow} z$$
$$T \cdot S: y \underset{T}{\longrightarrow} n \underset{S}{\longrightarrow} w$$

and if $x = y$, then $z = w$ so that $T \cdot S$ is a mapping. Also, if $x \neq y$, then $m \neq n$ because T is a 1–1 mapping; if $m \neq n$, then $z \neq w$ because S is a 1–1 mapping. Therefore, $T \cdot S$ is 1–1. To show that $T \cdot S$ is onto, we must show that if $a \in X$, then there is an element $x \in X$ such that $T \cdot S: x \longrightarrow a$. S is onto, so that there is an element $u \in X$ such that $S: u \longrightarrow a$. And since T is onto, there is an element $x \in X$ such that $T: x \longrightarrow u$. Then

$$T \cdot S: x \underset{T}{\longrightarrow} u \underset{S}{\longrightarrow} a \quad \text{or} \quad T \cdot S: x \longrightarrow a$$

Then $T \cdot S$ is onto. This concludes the proof that $T \cdot S \in M$.

(b) Define I as follows:

$$I: x \longrightarrow x \quad \text{for all } x \in X$$

Then $I \in M$, i.e., I is a 1–1 mapping of X onto X. If $T \in M$, consider $I \cdot T$ and $T \cdot I$. If for $x \in X$, we have $T: x \longrightarrow y$, then

$$I \cdot T: x \underset{I}{\longrightarrow} x \underset{T}{\longrightarrow} y \quad \text{or} \quad I \cdot T: x \longrightarrow y$$

and

$$T \cdot I: x \underset{T}{\longrightarrow} y \underset{I}{\longrightarrow} y \quad \text{or} \quad T \cdot I: x \longrightarrow y$$

In both cases we see that the effect of $I \cdot T$ and $T \cdot I$ on each element of X is exactly the same as the effect of T. Hence, we conclude

$$T = I \cdot T = T \cdot I$$

i.e., I is an identity element for M.

(c) If $T, S, U \in M$, we must show that

$$(T \cdot S) \cdot U = T \cdot (S \cdot U)$$

Suppose we have

$$T: x \longrightarrow y$$
$$S: y \longrightarrow z$$
$$U: z \longrightarrow w$$

Then by definition of $T \cdot S$ and $S \cdot U$ we have

$$T \cdot S: x \xrightarrow{T} y \xrightarrow{S} z \quad \text{or} \quad T \cdot S: x \longrightarrow z$$

and

$$S \cdot U: y \xrightarrow{S} z \xrightarrow{U} w \quad \text{or} \quad S \cdot U: y \longrightarrow w$$

Using the definition of $(T \cdot S) \cdot U$ and $T \cdot (S \cdot U)$ we have

$$(T \cdot S) \cdot U: x \xrightarrow{T \cdot S} z \xrightarrow{U} w \quad \text{or} \quad (T \cdot S) \cdot U: x \longrightarrow w$$

and

$$T \cdot (S \cdot U): x \xrightarrow{T} y \xrightarrow{S \cdot U} w \quad \text{or} \quad T \cdot (S \cdot U): x \longrightarrow w$$

Therefore, $(T \cdot S) \cdot U = T \cdot (S \cdot U)$ showing that the product of mappings is associative.

(d) Now show that every element $T \in M$ has an inverse in M. If $T: x \longrightarrow y$, define $T^{-1}: y \longrightarrow x$. T^{-1} is a mapping since if $y = w$ and $T: x \longrightarrow y$, $T: z \longrightarrow w$ then $x = z$ since T is 1-1. Therefore, if $y = w$ and $T^{-1}: y \longrightarrow x$ and $T^{-1}: w \longrightarrow z$, then $x = z$. The domain of T^{-1} is X since T is an onto mapping. Suppose that

$$T^{-1}: x \longrightarrow y \quad \text{and} \quad T^{-1}: u \longrightarrow v$$

and $y = v$. By definition of T^{-1} we have

$$T: y \longrightarrow x \quad \text{and} \quad T: v \longrightarrow u$$

Since T is a mapping and $y = v$, it follows that $x = u$. Thus T^{-1} is 1-1. If $w \in X$ we must show that there is $x \in X$ such that $T^{-1}: x \longrightarrow w$ to show that T^{-1} is onto. But for $w \in X$, T maps w into some element, say x, i.e., $T: w \longrightarrow x$ and by definition of T^{-1} it follows that $T^{-1}: x \longrightarrow w$. Hence, $T^{-1} \in M$ and it only remains to show that T^{-1} is the inverse of T. Suppose that $T: x \longrightarrow y$. Then

$$T \cdot T^{-1}: x \xrightarrow{T} y \xrightarrow{T^{-1}} x \quad \text{or} \quad T \cdot T^{-1}: x \longrightarrow x$$

and

$$T^{-1} \cdot T: y \xrightarrow{T^{-1}} x \xrightarrow{T} y \quad \text{or} \quad T^{-1} \cdot T: y \longrightarrow y$$

This shows that $T^{-1} \cdot T = I$ and $T \cdot T^{-1} = I$ and completes the proof of Theorem 28.

Although Theorem 28 makes no assumption about whether X is finite or infinite, we will be concerned only with finite sets. If X contains n elements, where $n \geq 1$, we denote by S_n the set of all 1-1 mappings of X onto X. The group $(S_n, \cdot)$ is called the *symmetric group* on n elements. We will refer to the symmetric group simply as S_n with the understanding that the binary operation is the product of mappings. It is common practice to refer to a group $(G, \cdot)$ as "the group G."

For the set $X = \{1, 2, 3\}$, S_3 consists of the six elements I, A, B, C, D, E given on p. 172. The group table is

·	I	A	B	C	D	E
I	I	A	B	C	D	E
A	A	I	E	D	C	B
B	B	D	I	E	A	C
C	C	E	D	I	B	A
D	D	B	C	A	E	I
E	E	C	A	B	I	D

When we speak of S_n we will always understand that the set in question is $\{1, 2, 3, \ldots, n\}$. The notation which we have used for elements of S_n, e.g.,

$$\begin{pmatrix} 1 & 2 & 3 & 4 & 5 \\ 2 & 1 & 4 & 3 & 5 \end{pmatrix}$$

in S_5, is rather cumbersome. Fortunately, there is a more convenient notation, called *cycle notation*. In S_5 the permutation

$$\begin{pmatrix} 1 & 2 & 3 & 4 & 5 \\ 3 & 4 & 5 & 1 & 2 \end{pmatrix}$$

is an example of a cycle. This may be written more compactly as

$$(1 \ 3 \ 5 \ 2 \ 4)$$

with the understanding that each element before the last is mapped into the element following and that the last element maps into the first. This may be pictured as follows:

$$(1 \longrightarrow 3 \longrightarrow 5 \longrightarrow 2 \longrightarrow 4)$$

A permutation such as

$$\begin{pmatrix} 1 & 2 & 3 & 4 & 5 \\ 1 & 4 & 3 & 2 & 5 \end{pmatrix}$$

is expressed as (24), it being understood that the missing elements are mapped into themselves. The permutation

$$\begin{pmatrix} 1 & 2 & 3 & 4 & 5 \\ 2 & 1 & 5 & 4 & 3 \end{pmatrix}$$

may be written as

$$\begin{pmatrix} 1 & 2 & 3 & 4 & 5 \\ 2 & 1 & 3 & 4 & 5 \end{pmatrix} \cdot \begin{pmatrix} 1 & 2 & 3 & 4 & 5 \\ 1 & 2 & 5 & 4 & 3 \end{pmatrix} = \begin{pmatrix} 1 & 2 & 3 & 4 & 5 \\ 2 & 1 & 5 & 4 & 3 \end{pmatrix}$$

In the cyclic notation

$$\begin{pmatrix} 1 & 2 & 3 & 4 & 5 \\ 2 & 1 & 3 & 4 & 5 \end{pmatrix} = (12)$$

and

$$\begin{pmatrix} 1 & 2 & 3 & 4 & 5 \\ 1 & 2 & 5 & 4 & 3 \end{pmatrix} = (35)$$

so that

$$\begin{pmatrix} 1 & 2 & 3 & 4 & 5 \\ 2 & 1 & 5 & 4 & 3 \end{pmatrix} = (12)(36)$$

In this notation the six elements of S_3 listed on p. 172 may be written as follows:

$$I, A = (23), B = (13), C = (12), D = (123), E = (132)$$

Observe that $(123) = (231) = (312)$. Generally, $(1234\ldots n) = (234\ldots n1) = (345\ldots n12) = (456\ldots n123)$, etc.

Consider the problem of forming products in this notation. Suppose we have A, B in S_5, where

$$A = \begin{pmatrix} 1 & 2 & 3 & 4 & 5 \\ 2 & 3 & 1 & 5 & 4 \end{pmatrix} \text{ and } B = \begin{pmatrix} 1 & 2 & 3 & 4 & 5 \\ 1 & 5 & 4 & 3 & 2 \end{pmatrix}$$

Then in cycle notation $A = (123)(45)$ and $B = (24)(34)$. We now wish to compute AB in cycle notation. *This amounts to determining the effect of AB on each of the elements 1, 2, 3, 4, 5 and expressing this in cycle notation.* Remember that here both A and B are expressed as products. What happens to 1 under $AB = (123)(45)(25)(34)$? To answer this we apply (123), (45), (25), (34) to 1 in succession:

$$1 \underset{(123)}{\longrightarrow} 2 \underset{(45)}{\longrightarrow} 2 \underset{(25)}{\longrightarrow} 5 \underset{(34)}{\longrightarrow} 5 \quad \text{or} \quad 1 \longrightarrow 5$$

In like manner we have

$$2 \underset{(123)}{\longrightarrow} 3 \underset{(45)}{\longrightarrow} 3 \underset{(25)}{\longrightarrow} 3 \underset{(34)}{\longrightarrow} 4 \quad \text{or} \quad 2 \longrightarrow 4$$

$$3 \underset{(123)}{\longrightarrow} 1 \underset{(45)}{\longrightarrow} 1 \underset{(25)}{\longrightarrow} 1 \underset{(34)}{\longrightarrow} 1 \quad \text{or} \quad 3 \longrightarrow 1$$

$$4 \underset{(123)}{\longrightarrow} 4 \underset{(45)}{\longrightarrow} 5 \underset{(25)}{\longrightarrow} 2 \underset{(34)}{\longrightarrow} 2 \quad \text{or} \quad 4 \longrightarrow 2$$

$$5 \underset{(123)}{\longrightarrow} 5 \underset{(45)}{\longrightarrow} 4 \underset{(25)}{\longrightarrow} 4 \underset{(34)}{\longrightarrow} 3 \quad \text{or} \quad 5 \longrightarrow 3$$

We conclude that
$$AB = (123)(45)(25)(34) = (153)(24)$$

It is no doubt clear that every permutation of S_n may be written as a product of cycles. Also notice that in the product, the cycles have no elements in common. For example, in (153)(24) the cycles (153) and (24) have no elements in common. This is described by saying that the cycles are *disjoint*.

EXAMPLE 4-10: We express the following permutation as a product of disjoint cycles:
$$\begin{pmatrix} 1 & 2 & 3 & 4 & 5 & 6 & 7 & 8 & 9 \\ 6 & 9 & 5 & 4 & 8 & 2 & 3 & 7 & 1 \end{pmatrix} = (1629)(3587)$$

EXAMPLE 4-11: We express the cycles of S_4 which involve only 3 elements:
$$(123), (124), (132), (134), (142), (143), (234), (243)$$

DEFINITION. A cycle involving m elements is called an m-*cycle*. A 2-cycle is called a *transposition*.

In studying the groups S_n, the 3-cycles and transpositions are very important as the following considerations will show. If $(n_1\, n_2\, n_3 \ldots n_k)$ is any cycle, we may write
$$(n_1\, n_2\, n_3 \ldots n_k) = (n_1\, n_2)(n_1\, n_3)(n_1\, n_4) \ldots (n_1\, n_k)$$

This shows that every cycle may be written as a product of transpositions. It then follows that every permutation may be written as a product of transpositions since every permutation may be written as a product of cycles. For example
$$(1629)(3587) = (16)(12)(19)(35)(38)(37)$$

It is always possible to write a permutation in many different ways as a product of transpositions. As an example, consider the following:
$$(235) = (23)(25)$$
$$= (15)(23)(15)(25)$$
$$= (24)(15)(24)(23)(15)(25)$$

Evidently, we could continue with the above equations to obtain even more expressions of (235) as products of transpositions. Observe that in each of the above expressions of (235) as a product of transpositions the number of transpositions is even.

Theorem 29. Let $A \in S_n$. If A can be written in one way as a product of an even number of transpositions, then every expression of A as a product of transpositions contains an even number of transpositions. If A can be written in one way as a product of an odd number of transpositions, then every expression of A as a product of transpositions contains an odd number of transpositions.

We only sketch a proof of this theorem. Complete proofs may be found in the following books:

Alexandroff, P.S., *An Introduction to the Theory of Groups*. New York: Hafner Publishing Co. Inc., 1959.

Papy, Georges, *Groups*. New York: St. Martin's Press, Inc., 1964.

Sketch of Proof. For the proof we form all the ordered pairs (i, j) with $i < j$ and $i, j \in \{1, 2, 3, \ldots, n\}$. We may list these as follows: $(1, 2), (1, 3), (1, 4), \ldots, (1, n)$, $(2, 3), (2, 4), \ldots, (2, n) \ldots, (n-1, n)$. Now let A be expressed as a product of transpositions. If a transposition (i, j) occurs in A an even number of times (note that this includes the case when (ij) does not occur as a factor of A, i.e., (ij) occurs $0 = 2 \cdot 0$ times), then under the pair (i, j) of the above list we record 1; if (ij) occurs in A an odd number of times, we record -1. As an example of this consider $(235) \in S_5$. The above pairs in this case are $(1, 2), (1, 3), (1, 4), (1, 5), (2, 3), (2, 4), (2, 5)$, $(3, 4), (3, 5), (4, 5)$. If we write

$$(235) = (23)(25)$$

then we record a 1 under every pair above except under $(2, 3)$ and $(2, 5)$; under these we record a -1. (235) may be written, for example, as

(a) $(23)(25)$
(b) $(15)(23)(15)(25)$
(c) $(24)(15)(24)(23)(15)(25)$

For (a) we obtain $1, 1, 1, 1, -1, 1, -1, 1, 1, 1$; for (b) we obtain $1, 1, 1, 1, -1, 1, -1$, $1, 1, 1$; for (c) we obtain $1, 1, 1, 1, -1, 1, -1, 1, 1, 1$. To continue the proof, when 1's and -1's have been recorded we multiply all these, obtaining as a product either 1 or -1. If an even number of -1's is recorded we get a product of 1 and if an odd number of -1's is recorded we obtain a product of -1. Evidently, the permutation A cannot produce in this way both 1 and -1. Therefore, A cannot be expressed as a product of an even number of transpositions and also an odd number of transpositions. This completes our sketch of the proof.

DEFINITION. A permutation A is called *even* or *odd* depending upon its expression as a product of an even or odd number of transpositions. The subset of S_n consisting of all even permutations is denoted by A_n.

Theorem 30. A_n is a subgroup of S_n.

Proof. (a) If $A, B \in A_n$, then A and B can each be written as a product of an even number of transpositions, say s, t respectively. Since s, t are each even, $s + t$ is even and AB can be written as $s + t$ transpositions, an even number. Hence, $AB \in A_n$.

(b) The identity I may be written as $I = (12)(12)$. Hence, I is even and $I \in A_n$.

(c) For $A \in A_n$ if A^{-1} is an odd permutation, then AA^{-1} is an odd permutation, contrary to the fact that $AA^{-1} = I$ is an even permutation.

EXAMPLE 4-12: The elements of S_3 are

$$I, (12), (13), (23), (123), (132)$$

The even permutations of S_3 are

$$I, (123) = (12)(13), (132) = (13)(12)$$

Hence, $A_3 = \{I, (123), (132)\}$.

DEFINITION. If G is a group containing an infinite number of elements, G is said to have *infinite order*. If G has a finite number of elements, say n, then G is said to have *order n*.

EXAMPLE 4-13: S_3 has order 6 and A_3 has order 3. The group of all rational numbers with binary operation addition has infinite order.

Theorem 31. The order of S_n is $n!$ and the order of A_n is $n!/2$.

Proof. The notation $n!$ means the number $1 \cdot 2 \cdot 3 \ldots n$. To determine the order of S_n we need to calculate the number of 1-1 mappings of $\{1, 2, 3, \ldots, n\}$ onto $\{1, 2, 3, \ldots, n\}$. Each of these mappings may be expressed as follows:

$$\begin{pmatrix} 1 & 2 & 3 & 4 & \cdots & n \\ a_1 & a_2 & a_3 & a_4 & \cdots & a_n \end{pmatrix}$$

where each of the elements $a_1, a_2, a_3, \ldots, a_n$ are members of $\{1, 2, 3, \ldots, n\}$. Since the mapping is 1-1 the a's must be distinct, i.e., $a_i \neq a_j$ if $i \neq j$. Hence, a_1 may be chosen in n ways, a_2 in $n - 1$ ways, a_3 in $n - 2$ ways, a_4 in $n - 3$ ways, $\ldots$, a_n in 1 way. Therefore, there are $n \cdot (n - 1) \cdot (n - 2) \cdot (n - 3) \ldots 2 \cdot 1 = n!$ 1-1 mappings of $\{1, 2, 3, \ldots, n\}$ onto $\{1, 2, 3, \ldots, n\}$. It then follows that S_n has order $n!$.

Now let us define a mapping T of A_n into the set of odd permutations as follows:

$$T: E \longrightarrow E(12)$$

for all $E \in A_n$. Since E is even, $E(12)$ is odd. T is clearly a mapping. Also, T is 1-1 and onto. Suppose that $E \in A_n$, $D \in A_n$ and that $E(12) = D(12)$. Then $E(12)(12) = D(12)(12)$ or $E = D$. Hence, T is 1-1. To show that T is onto the set of odd permutations, let Q be any odd permutation. The equation

$$Q = X(12)$$

has a solution by Theorem 22. The solution X is unique and X must be even since Q is odd. Then by definition of T we have

$$T: X \longrightarrow X(12) = Q$$

which shows that T is onto. This demonstrates that there are exactly as many even permutations of S_n as odd permutations since each even permutation is paired by T with exactly one odd permutation. It then follows that $n!/2$ is the number of even permutations and also the number of odd permutations. Thus the order of A_n is $n!/2$ and this completes the proof.

Example 4-14: Compute the subgroup of S_4 generated by (23) and (234), i.e., compute [(23), (234)].

Solution: A routine way of doing this is to use a table. Since the subgroup [(23), (234)] must contain the elements I, (23), (234), we start with the table shown below:

	I	(23)	(234)
I	I	(23)	(234)
(23)	(23)	I	(24)
(234)	(234)	(34)	(243)

We start with this table because we know that the subgroup [(23), (234)] must contain all the products given in this table. The products obtained in the table are I, (23), (234), (24), (34), (243); the elements (24), (34), (243) are "new" elements which we did not have before. Now expand the table to include these new elements. Notice that in the table below the original table is in the upper left-hand corner.

	I	(23)	(234)	(24)	(34)	(243)
I	I	(23)	(234)	(24)	(34)	(243)
(23)	(23)	I	(24)	(234)	(243)	(34)
(234)	(234)	(34)	(243)	(23)	(24)	I
(24)	(24)	(243)	(34)	I	(234)	(23)
(34)	(34)	(234)	(23)	(243)	I	(24)
(243)	(243)	(24)	I	(34)	(23)	(234)

Also, no "new" elements have been produced in this table, and each element has an inverse. Therefore, this is the table for the group [(23), (234)] so that [(23), (234)] = {I, (23), (234), (24), (34) (243)}. We might call this process of computing (23), (234) the "expanding table" method.

Exercises

63. In S_6, write the following as products of disjoint cycles.

(a) $\begin{pmatrix} 1 & 2 & 3 & 4 & 5 & 6 \\ 6 & 1 & 5 & 2 & 3 & 4 \end{pmatrix}$ = _____

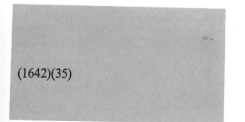

(1642)(35)

(145)(263)

(12)(45)

(146)

(143628597)

(1863)(495)

(18)(29)(37)(56)

(14)(235)(6987)

(129756)

(35764)

(15)(34)

(13)(276)

(13)(24)

(1432)

I

(15)

(34)

(264)

(b) $\begin{pmatrix} 1 & 2 & 3 & 4 & 5 & 6 \\ 4 & 6 & 2 & 5 & 1 & 3 \end{pmatrix} = $ _____

(c) $\begin{pmatrix} 1 & 2 & 3 & 4 & 5 & 6 \\ 2 & 1 & 3 & 5 & 4 & 6 \end{pmatrix} = $ _____

(d) $\begin{pmatrix} 1 & 2 & 3 & 4 & 5 & 6 \\ 4 & 2 & 3 & 6 & 5 & 1 \end{pmatrix} = $ _____

64. In S_9, write the following as products of disjoint cycles.

(a) $\begin{pmatrix} 1 & 2 & 3 & 4 & 5 & 6 & 7 & 8 & 9 \\ 4 & 8 & 6 & 3 & 9 & 2 & 1 & 5 & 7 \end{pmatrix} = $ _____

(b) $\begin{pmatrix} 1 & 2 & 3 & 4 & 5 & 6 & 7 & 8 & 9 \\ 8 & 2 & 1 & 9 & 4 & 3 & 7 & 6 & 5 \end{pmatrix} = $ _____

(c) $\begin{pmatrix} 1 & 2 & 3 & 4 & 5 & 6 & 7 & 8 & 9 \\ 8 & 9 & 7 & 4 & 6 & 5 & 3 & 1 & 2 \end{pmatrix} = $ _____

(d) $\begin{pmatrix} 1 & 2 & 3 & 4 & 5 & 6 & 7 & 8 & 9 \\ 4 & 3 & 5 & 1 & 2 & 9 & 6 & 7 & 8 \end{pmatrix} = $ _____

65. Compute each of the following as products of disjoint cycles.

(a) $(156)(2975) = $ _____

(b) $(34)(75)(456) = $ _____

(c) $(1542)(12)(341) = $ _____

(d) $(6253)(76)(1357) = $ _____

(e) $(1234)(1234) = $ _____

(f) $(1234)^3 = (1234)^2(1234) = $ _____

(g) $(1234)^4 = (1234)^3(1234) = $ _____

(h) $(12345)(12)(54321) = $ _____

(i) $(34)^{-1} = $ _____

(j) $(246)^{-1} = $ _____

(4321)
(54321)
(14)(23)
(23)
(13)(24)
(34)

2

(acb)
I
3
3

(ac)(bd)
(adcb)
I
4
4

(ced)
(ab)
(cde)
(ab)(ced)
I
6

6

(k) $(1234)^{-1} =$ _____

(l) $(12345)^{-1} =$ _____

(m) $(123)^{-1}(12)(1234) =$ _____

(n) $(1234)^{-1}(12)(1234) =$ _____

(o) $(1234)^{-2} = (1234)^{-1}(1234)^{-1} =$ _____

(p) $(1234)^{-2}(12)(1234)^2 =$ _____

66. The order of a transposition (ab) is _____

67. (a) Compute: $(abc)^2 =$ _____
    (b) Compute: $(abc)^3 = (abc)^2(abc) =$ _____
    (c) The order of (abc) is _____
    (d) The order of any 3-cycle is _____

68. (a) Compute: $(abcd)^2 =$ _____
    (b) Compute: $(abcd)^3 =$ _____
    (c) Compute: $(abcd)^4 =$ _____
    (d) The order of $(abcd)$ is _____
    and, hence, the order of any 4-cycle is _____

69. (a) Compute: $[(ab)(cde)]^2 = (ab)(cde)(ab)(cde) =$ _____
    (b) Compute: $[(ab)(cde)]^3 =$ _____
    (c) Compute: $[(ab)(cde)]^4 =$ _____
    (d) Compute: $[(ab)(cde)]^5 =$ _____
    (e) Compute: $[(ab)(cde)]^6 =$ _____
    (f) The order of $(ab)(cde)$ is _____
    (g) If a 2-cycle and a 3-cycle are disjoint, then their product has order _____

182 Groups

2

$\{I, (12), (13)\}$

(123)

3

I

(123), (132)

$I, (12), (13), (123), (132)$

(13)
(23)
(23)
(12)
(23)

70. In S_4, compute the subgroup generated by (12) and (13), i.e., compute [(12), (13)]. *Solution.* Since [(12), (13)] is a group, it must contain all the powers of both (12) and (13). But since each of these elements are transpositions and have order _____, the powers of these elements result in the set _____. Also, since [(12), (13)] is a group, $(12) \cdot (13) \in [(12), (13)]$, i.e., $(12) \cdot (13) = $ _____ $\in [(12), (13)]$. Then powers of (123) must be in [(12), (13)]. Since (123) has order _____, the powers are just $(123)^0$, $(123)^1$, $(123)^2$ or $(123)^0 = $ _____, $(123)^1 = $ _____ and $(123)^2 = $ _____. So far the subgroup [(12), (13)] contains the elements _____.
Next, the subgroup must contain all the products of these elements; products of the elements in the sets $\{I, (12), (13)\}$ and $\{(123), (132)\}$ do not result in any "new" elements. Now we may check these products:

(12)(123), (12)(132), (13)(123), (13)(132), (123)(12), (132)(12), (123)(13), (32)(13).

(12)(123) = _____
(12)(132) = _____
(13)(123) = _____
(13)(132) = _____
(123)(12) = _____

Groups 183

(13)

(12)

(23)

(23)

I, (12), (13), (23), (123), (132)

(132)(12) = _____

(123)(13) = _____

(132)(13) = _____

The results of these products produce only the one new element _____.

Then the following elements are in the subgroup [(12), (13)]: _____

Construct a table below for these elements:

	I	(12)	(13)	(23)	(123)	(132)
I						
(12)						
(13)						
(23)						
(123)						
(132)						

	I	(12)	(13)	(23)	(123)	(132)
I	I	(12)	(13)	(23)	(123)	(132)
(12)	(12)	I	(123)	(132)	(13)	(23)
(13)	(13)	(132)	I	(123)	(23)	(12)
(23)	(23)	(123)	(132)	I	(12)	(13)
(123)	(123)	(23)	(12)	(13)	(132)	I
(132)	(132)	(13)	(23)	(12)	I	(123)

It is clear from the table that $\{I,$ (12), (13), (23), (123), (132)$\}$ is a group which contains (12) and (13). Also, it is clear from our calculations that no

"smaller" subgroup of S_4 contains (12) and (13). Hence, $[(12), (13)] = \{I, (12), (13), (23), (123), (132)\}$.

71. Compute $[(12), (13)]$ by the "expanding table" method.

72. Compute $[(123), (234)]$ by the "expanding table" method.

73. Determine the elements of A_4 in cycle notation.

74. Let H be a subgroup of S_n. If $x, y \in \{1, 2, 3, 4, \ldots, n\}$, define xRy to mean that there is an element $A \in H$ such that $A\colon x \longrightarrow y$. Show that R is an equivalence relation on the set $\{1, 2, 3, 4, \ldots, n\}$.
(a) For the subgroup $\{I, (12)\}$ of S_3, compute R and the equivalence classes $\bar{1}, \bar{2}, \bar{3}$ associated with R.
(b) For the subgroup $\{I, (23), (234), (24), (34), (243)\}$ of S_4, compute R and the corresponding equivalence classes.

4.6 Cosets

Let G be a group and let H be a subgroup of G. A relation on G may be defined as follows: for $a, b \in G$

$$aRb \text{ if and only if } a^{-1}b \in H$$

(a) *R is reflexive.* This is true because $e \in H$ and for any $g \in G$, $g^{-1}g = e \in H$.
(b) *R is symmetric.* If aRb, the definition of R requires that $a^{-1}b \in H$. Then, since H is a subgroup of G, $(a^{-1}b)^{-1} \in H$. But $(a^{-1}b)^{-1} = b^{-1}a$. Hence, $b^{-1}a \in H$ so that bRa. Thus R is symmetric.
(c) *R is transitive.* If aRb and bRc, we have $a^{-1}b \in H$ and $b^{-1}c \in H$. Since H is a subgroup of G, we have $(a^{-1}b)(b^{-1}c) \in H$, i.e., $(a^{-1}b)(b^{-1}c) = a^{-1}(bb^{-1})c = a^{-1} \cdot e \cdot c = a^{-1}c \in H$. This implies that aRc.

From the above arguments we may conclude that R is an equivalence relation on G. It then follows that there is a partition of G associated with R. If $g \in G$, the partition of G consists of the sets $\bar{g} = \{x \mid gRx \text{ and } x \in G\}$. These equivalence classes associated with the equivalence relation R and the subgroup H are called the *left cosets* of H in G.

Theorem 32. Let H be a subgroup of the group G. Then the left cosets of H in G may be described as follows: for each $g \in G$
$$\bar{g} = \{gh \mid h \in H\}$$
Proof. We denote $\{gh \mid h \in H\}$ by gH. The theorem may then be proved by showing that $\bar{g} \subseteq gH$ and $gH \subseteq \bar{g}$.

(i) $\bar{g} \subseteq gH$. Suppose that $x \in \bar{g}$. Then by definition of $\bar{g}$, gRx and $g^{-1}x \in H$. Therefore, $g^{-1}x = h$, where $h \in H$. Hence, $x = gh$ and this implies that $x \in gH$. This shows that $\bar{g} \subseteq gH$.

(ii) $gH \subseteq \bar{g}$. If $x \in gH$, then $x = gh$ for some $h \in H$. From $x = gh$ it follows that $g^{-1}x = h \in H$ or gRx. But with gRx we have $x \in \bar{g}$. Thus $gH \subseteq \bar{g}$. This completes the proof.

EXAMPLE 4-15: In $S_3 = \{I, (12), (13), (23), (123), (132)\}$ we compute the left cosets of the subgroup $H = \{I, (12)\}$.

$$IH = \{I \cdot h \mid h \in H\} = \{h \mid h \in H\} = H$$
$$(12)H = \{(12)h \mid h \in H\} = \{(12) \cdot I, (12)(12)\} = H$$
$$(13)H = \{(13)h \mid h \in H\} = \{(13) \cdot I, (13)(12)\} = \{(13), (132)\}$$
$$(23)H = \{(23)h \mid h \in H\} = \{(23) \cdot I, (23) \cdot (12)\} = \{(23), (123)\}$$
$$(123)H = \{(123)h \mid h \in H\} = \{(123) \cdot I, (123) \cdot (12)\} = \{(123), (23)\}$$
$$(132)H = \{(132)h \mid h \in H\} = \{(132) \cdot I, (132) \cdot (12)\} = \{(132), (13)\}$$

Here there are only 3 distinct, left cosets: H, $(13)H$, $(23)H$.

EXAMPLE 4-16: Let H be the subgroup of $(Z, +)$ of all multiples of 2. The left cosets of H are as follows:

$$0H = \{0 + h \mid h \in H\} = H$$
$$1H = \{1 + h \mid h \in H\} = \text{the set of odd integers}$$

In this case there are only two distinct left cosets. To see this, remember that a left coset nH is the equivalence class $[n]$ and that $[n] = [m]$ if and only if nRm which means that in this case $-n + m \in H$ or that $n \equiv m(2)$. Since for any $n \in Z$ we have $n \equiv 0(2)$ or $n \equiv 1(2)$, $[n] = [0]$ or $[n] = [1]$, i.e., $nH = 0H$ or $nH = 1H$.

Let G be a group and let H be a subgroup of G. If $g \in G$, we may define a mapping T from H into the left coset gH as follows:

$$T: h \longrightarrow gh \quad \text{for all } h \in H$$

Lemma. The mapping T is a 1-1 mapping of H onto gH.

Proof. T is a mapping since $h_1 = h_2$ implies $gh_1 = gh_2$. If $gh_1 = gh_2$, then $g^{-1}(gh_1) = g^{-1}(gh_2)$ or $h_1 = h_2$. Hence, T is 1-1. If $gh \in gH$, then h maps onto hg so that T is onto.

In case the subgroup H has finite order, the lemma above allows us to conclude that each left coset has the same number of elements as H. Notice that this is the case in example 4–15.

Theorem 33 (*Lagrange*). Let G be a group of finite order n and let H be a subgroup of G. Then if t is the order of H, $t \mid n$.

Proof. The left cosets of H constitute a partition of G so that we may write

$$G = g_1 H \cup g_2 H \cup \cdots \cup g_k H$$

where $g_1 H, g_2 H, g_3 H, \ldots, g_k H$ are all the distinct left cosets of H. We also know, since the left cosets form a partition of G, that two distinct cosets are disjoint. Therefore, we may add the number of elements of the cosets and obtain the order of G. But by the lemma above, each coset has the same number of elements as H, namely t. Hence,

$$n = \underbrace{t + t + t + \cdots + t}_{k \text{ summands of } t}$$
$$= k \cdot t$$

Thus $t \mid n$ and the theorem is proved.

EXAMPLE 4-17: Suppose a group G has order 12. Then since $12 = 1 \cdot 2 \cdot 2 \cdot 3$, the only possible orders for subgroups are 1, 2, 3, 4, 6, 12. Such a group, for example, cannot have a subgroup of order 5.

Theorem 34. If G is a group of prime order, then G is a cyclic group.

Proof. Since G has prime order p and $p > 1$, G contains some element $g \neq e$. Then the cyclic subgroup $[g] \neq \{e\}$. The only factors of p are 1 and p so that the only possible order for $[g]$ is 1 and p. The order of $[g]$ is not 1 since $[g] \neq \{e\}$. Hence, the order of $[g]$ is p, so $[g] = G$.

Theorem 35. Let G be a group of finite order n. If $g \in G$, then the order of g divides n.

Proof. If t is the order of the element g, then $[g]$ has order t. Since $[g]$ is a subgroup of G, $t \mid n$ by the theorem of Lagrange.

Theorem 36 (*Fermat*). Let a be a positive integer and let p be a prime. Then

$$a^p \equiv a \ (p)$$

Proof. Since a is a positive integer, $\bar{a} \in Z_p^*$. In the group $(Z_p^*, \otimes)$ the element $\bar{a}$ has an order, say t, which divides the order of Z_p^* by Theorem 35. $Z_p^* = \{\bar{1}, \bar{2}, \bar{3}, \ldots, \overline{(p-1)}\}$ so that the order of Z_p^* is $p - 1$. Then $p - 1 = t \cdot k$ for some integer k. The order of $\bar{a}$ is t so that

$$\bar{a}^t = \bar{1}$$

or
$$\underbrace{\bar{a} \otimes \bar{a} \otimes \bar{a} \otimes \cdots \otimes \bar{a}}_{t \text{ factors of } \bar{a}} = \bar{1}$$

Hence, $\bar{a}^t = \bar{1}$. Next, $(\bar{a}^t)^k = \bar{1}^k = \bar{1}$ and
$$\bar{a}^{tk} = \bar{1}$$

It then follows that $a^{tk} \equiv 1\,(p)$ or that $a^{p-1} \equiv 1\,(p)$ since $p - 1 = tk$. The congruence $a^{p-1} \equiv 1\,(p)$ may be multiplied by a to give
$$a \cdot a^{p-1} \equiv a\,(p)$$
or
$$a^p \equiv a\,(p)$$

This completes the proof.

If H is a subgroup of a group G, a *right coset* of H is a set $Hg = \{hg \mid h \in H\}$. It is easy to show that the number of elements of any left coset is the same as the number of elements of H. This may be done in the same way that the corresponding result for left cosets is proved. We may also prove that the number of right cosets of H and the number of left cosets of H is the same. To do this, define a mapping Q as follows:

$$Q: \quad Hg \longrightarrow g^{-1}H$$

To show that Q is a mapping, suppose that $Hg_1 = Hg_2$ for $g_1, g_2 \in G$. Then we wish to show that $g_1^{-1}H = g_2^{-1}H$. If $x \in g_1^{-1}H$, then $x = g_1^{-1}h$ for some $h \in H$. From $x = g_1^{-1}h$, $x^{-1} = (g_1^{-1}h)^{-1} = h^{-1}g_1 \in Hg_1$. But $Hg_1 = Hg_2$, which implies that there is an $h_1 \in H$ such that $x^{-1} = h^{-1}g_1 = h_1g_2$. From $x^{-1} = h_1g_2$ it follows that $x = (x^{-1})^{-1} = (h_1g_2)^{-1} = g_2^{-1}h_1^{-1} \in g_2^{-1}H$. Therefore, if $x \in g_1^{-1}H$, then $x \in g_2^{-1}H$ and $g_1^{-1}H \subseteq g_2^{-1}H$. In a similar manner it is possible to show that $g_2^{-1}H \subseteq g_1^{-1}H$ so that $g_1^{-1}H = g_2^{-1}H$; Q is thus a mapping. Also, a similar argument shows that Q is 1–1; it is clearly onto.

DEFINITION. The number of right cosets (or left cosets) of a subgroup H in a group G is called the *index* of H in G.

By the theorem of Lagrange, if n is the order of a group G and t is the order of a subgroup H of G, then $n = t \cdot k$ for some integer k. Then k is the index of H in G. The index of H in G is denoted by $[G:H]$.

Example 4-18: The subgroup $H = \{I, (12)\}$ of S_3 has index 3.

Exercises

$\{I, (123), (132)\}$

75. Compute the right cosets of A_3 in S_3.

    (a) $A_3 = $ _____

$\{I, (123), (132)\}$

$\{(123), (132), I\}$

$\{(132), I, (123)\}$

$\{(12), (23), (13)\}$

$\{(13), (12), (23)\}$

$\{(23), (13), (12)\}$

2

3

2

H, H

H, H

$\{(13), (123)\}$, $\{(13), (132)\}$

$\{(23), (132)\}$, $\{(23), (123)\}$

$\{(123), (13)\}$, $\{(123), (23)\}$

$\{(132), (23)\}$, $\{(132), (13)\}$

$H, H(13), H(23)$

$H, (13)H, (23)H$

No

Because $(123) \neq (132)$

No

Because $(123) \neq (132)$

2

3

(b) $A_3 I =$ ____________

(c) $A_3 \cdot (123) =$ ____________

(d) $A_3 \cdot (132) =$ ____________

(e) $A_3 \cdot (12) =$ ____________

(f) $A_3 \cdot (13) =$ ____________

(g) $A_3 \cdot (23) =$ ____________

(h) How many left and right cosets of A_3 are there? ____________

(i) The order of A_3 is ____________ and the index of A_3 in S_3 is ____________

76. Compute the left and right cosets of $H = \{I, (12)\}$ in S_3.

(a) $HI =$ ________, $IH =$ ________

(b) $H(12) =$ ________, $(12)H =$ ________

(c) $H(13) =$ ________, $(13)H =$ ________

(d) $H(23) =$ ________, $(23)H =$ ________

(e) $H(123) =$ ________, $(123)H =$ ________

(f) $H(132) =$ ________, $(132)H =$ ________

(g) What are the distinct right cosets of H?

____

(h) What are the distinct left cosets of H?

____

(i) Is it true that $H(13) = (13)H$? ________

Why? ____________

(j) Is it true that $H(23) = (23)H$? ________

Why? ____________

(k) The order of H is ____________

(l) The index of H in S_3 is ____________

Groups 189

77. Compute the right cosets of $H = [\bar{2}]$ in $(Z_{10}, \oplus)$.

(a) $[\bar{2}] = $ _____ $\{\bar{0}, \bar{2}, \bar{4}, \bar{6}, \bar{8}\}$

(b) $H\bar{0} = $ _____ H , $H\bar{2} = $ _____ H , $H\bar{4} = $ _____ H

$H\bar{6} = $ _____ H , $H\bar{8} = $ _____ H

(c) $H\bar{1} = $ _____ $\{\bar{1}, \bar{3}, \bar{5}, \bar{7}, \bar{9}\}$

$H\bar{3} = $ _____ $\{\bar{3}, \bar{5}, \bar{7}, \bar{9}, \bar{1}\}$

$H\bar{5} = $ _____ $\{\bar{5}, \bar{7}, \bar{9}, \bar{1}, \bar{3}\}$

$H\bar{7} = $ _____ $\{\bar{7}, \bar{9}, \bar{1}, \bar{3}, \bar{5}\}$

$H\bar{9} = $ _____ $\{\bar{9}, \bar{1}, \bar{3}, \bar{5}, \bar{7}\}$

(d) What are the distinct right cosets of H?

_____ $H, H\bar{1}$

(e) The order of H is _____ 5

(f) The index of H in $(Z_{10}, \oplus)$ is _____ 2

78. Compute the right cosets of the subgroup $H = [3]$ in $(Z, +)$. Notice here that since the binary operation is $+$, a positive power of 3, say 3^n, is $3 + 3 + 3 + \cdots + 3$ or $n \cdot 3$. Therefore, since $[3]$ is the subgroup of all powers of 3, $[3]$ is the subgroup of all multiples of 3.

(a) $H = [3] = \{n \cdot 3 \mid n \in Z\}$.

(b) $H0 = $ _____ $\{h + 0 \mid h \in H\} = H$

(c) If $n \in H$, then $Hn = $ _____ $\{h + n \mid h \in H\} = H$

(d) $H1 = $ _____ $\{h + 1 \mid h \in H\}$

(e) If $n = 3q + 1$, then $Hn = $ _____ $\{(h + 3q) + 1 \mid h \in H\} \subseteq H1$

(f) $H2 = $ _____ $\{h + 2 \mid h \in H\}$

(g) If $n = 3q + 2$, then $Hn = $ _____ $\{(h + 3q) + 2 \mid h \in H\} \subseteq H2$

(h) For any $n \in Z$, we may write

$n = 3q + r, 0 \le r < 3$ for some $q, r \in Z$

190   Groups

Hence, $n - r = 3q \in H$ and this implies by the definition of R on p. 184 that rRn. Then $\bar{n} = \bar{r}$ or, in this case, $Hn = Hr$. Therefore, every coset of H is H(if $r = 0$), $H1$ (if $r = 1$) or $H2$ (if $r = 2$).

(i) The index of H in Z is _____

3

$\{h + 0 \mid h \in H\} = H$
$\{h + 1 \mid h \in H\}$
$\{h + 2 \mid h \in H\}$
$\{h + 3 \mid h \in H\}$
$\{h + 4 \mid h \in H\}$
$\{h + 5 \mid h \in H\}$
H
$H1$
$H2$
$H3$
$H4$
$H5$

6

6

Yes. Z and H here are an example of this.

79. Compute the right cosets of the subgroup $H = [6]$ of $(Z, +)$.

(a) $H = [6] = \{n \cdot 6 \mid n \in Z\}$
(b) $H0 =$ _____
(c) $H1 =$ _____
(d) $H2 =$ _____
(e) $H3 =$ _____
(f) $H4 =$ _____
(g) $H5 =$ _____
(h) If $n \in Z$ and $n = 6q$, then $Hn =$ _____
(i) If $n \in Z$ and $n = 6q + 1$, then $Hn =$ _____
(j) If $n \in Z$ and $n = 6q + 2$, then $Hn =$ _____
(k) If $n \in Z$ and $n = 6q + 3$, then $Hn =$ _____
(l) If $n \in Z$ and $n = 6q + 4$, then $Hn =$ _____
(m) If $n \in Z$ and $n = 6q + 5$, then $Hn =$ _____
(n) How many distinct cosets of H are there in Z? _____

(o) What is the index of H in Z, i.e., what is $[Z : H]$? _____

(p) Is it possible for a group G to have infinite order and for G to have a subgroup H of finite index in G? _____

Groups   191

80. Make the group table below for $(P_9, \otimes)$. Remember that P_9 consists of all $\bar{r}$ such that $0 < r < 9$ and $GCD(r, 9) = 1$. Then $P_9 = $ _____

$\{\bar{1}, \bar{2}, \bar{4}, \bar{5}, \bar{7}, \bar{8}\}$

$\otimes$	$\bar{1}$	$\bar{2}$	$\bar{4}$	$\bar{5}$	$\bar{7}$	$\bar{8}$
$\bar{1}$	$\bar{1}$	$\bar{2}$	$\bar{4}$	$\bar{5}$	$\bar{7}$	$\bar{8}$
$\bar{2}$	$\bar{2}$	$\bar{4}$	$\bar{8}$	$\bar{1}$	$\bar{5}$	$\bar{7}$
$\bar{4}$	$\bar{4}$	$\bar{8}$	$\bar{7}$	$\bar{2}$	$\bar{1}$	$\bar{5}$
$\bar{5}$	$\bar{5}$	$\bar{1}$	$\bar{2}$	$\bar{7}$	$\bar{8}$	$\bar{4}$
$\bar{7}$	$\bar{7}$	$\bar{5}$	$\bar{1}$	$\bar{8}$	$\bar{4}$	$\bar{2}$
$\bar{8}$	$\bar{8}$	$\bar{7}$	$\bar{5}$	$\bar{4}$	$\bar{2}$	$\bar{1}$

(a) $[\bar{1}] = $ _____ , (b) $[\bar{2}] = $ _____

$\{\bar{1}\}, \{\bar{1}, \bar{2}, \bar{4}, \bar{8}, \bar{7}, \bar{5}\}$

(c) $[\bar{4}] = $ _____ , (d) $[\bar{5}] = $ _____

$\{\bar{1}, \bar{4}, \bar{7}\}, \{\bar{1}, \bar{5}, \bar{7}, \bar{8}, \bar{4}, \bar{2}\}$

(e) $[\bar{7}] = $ _____ , (f) $[\bar{8}] = $ _____

$\{\bar{1}, \bar{7}, \bar{4}\}, \{\bar{1}, \bar{8}\}$

(g) Is $(P_9, \otimes)$ a cyclic group? _____

Yes

What are the single elements which generate P_9?

_____

$\bar{2}$ and $\bar{5}$

(h) Are the subgroups listed in (a)—(f) the only subgroups of P_9? _____

Yes

Why? _____

Cyclic groups have only cyclic subgroups.

1

(i) The order of $[\bar{1}]$ is _____ and $[P_9 : [\bar{1}]] = $ _____

6

6

(j) The order of $[\bar{2}]$ is _____ and $[P_9 : [\bar{4}]] = $ _____

1

3

(k) The order of $[\bar{4}]$ is _____ and $[P_9 : [\bar{4}]] = $ _____

2

6

(l) The order of $[\bar{5}]$ is _____ and

192 Groups

1	$[P_9 : [\bar{5}]] = $ _____
3	(m) The order of $[\bar{7}]$ is _____ and
2	$[P_9 : [\bar{7}]] = $ _____
2	(n) The order of $[\bar{8}]$ is _____ and
3	$[P_9 : [\bar{8}]] = $ _____
	(o) If $H = [\bar{2}]$, what is the only left coset of H
H	in P_9? _____
	(p) If $H = [\bar{4}]$, what are the 2 right cosets of H
$H, H\bar{2} = H\bar{5} = H\bar{8}$	in P_9? _____
	(q) If $H = [\bar{5}]$, what is the only right coset of H
H	in P_9? _____
	(r) If $H = [\bar{7}]$, what are the 2 right cosets of H
$H, H\bar{2} = H\bar{5} = H\bar{8}$	in P_9? _____
	(s) If $H = [\bar{8}]$, what are the 3 right cosets of H
$H, H\bar{2} = H\bar{7}, H\bar{4} = H\bar{5}$	in P_9? _____
6	(t) Since $(P_9, \otimes)$ has order _____,
	the possible orders for subgroups H of P_9 are
1, 2, 3, 6	_____. Does P_9 have a subgroup
Yes	of each of these possible orders? _____

81. If G is a group and H is a subgroup of G, what can you conclude about H if a coset Hg contains 14 elements? _____

H has order 14

If n is the order of G, what can you conclude? _____

$14 \mid n$

If $n = 14 \cdot k$ where k is some positive integer, $[G : H] = $ _____

k

82. If G is a group of order 15, what are the possible orders for subgroups of G? _____

1, 3, 5, 15

What are the possible indices for subgroups of G? _____ 1, 3, 5, 15 _____. If H is a subgroup of G whose order is known to be greater than 5, what can you conclude about H? ___ $H = G$ ___

83. If G is a group of order 32, what are the possible orders for subgroups of G? ___ 1, 2, 4, 8, 16, 32 ___
What are the possible indices for subgroups of G? _____ 1, 2, 4, 8, 16, 32 _____. If H is a subgroup of G which is known to have order greater than 4, what can you conclude about H? ___ H has order 8, 16, or 32 ___

84. Let G be a group of order n.

   (a) If $g \in G$ and g has order 5, then we may conclude that ___ $5 \mid n$ ___

   (b) If $g \in G$ and g has order 7, then we may conclude that ___ $7 \mid n$ ___

   (c) If G contains two elements, one of order 5 and one of order 7, and if the order of G is less than 40, what can you conclude about the order of G? ___ G has order 35 ___

85. Let H and K be subgroups of a group G and suppose that H has order t and K has order s. Prove that if $\text{GCD}(t, s) = 1$, then $H \cap K = \{e\}$.

86. Make a group table for $(P_{21}, \otimes)$ and compute all the subgroups.

87. Suppose that H is a subgroup of a group G and that $Hg_1, Hg_2, \ldots, Hg_k$ are all the distinct right cosets of H in G, i.e.,
$$G = Hg_1 \cup Hg_2 \ldots \cup Hg_k$$

Show that $g_1^{-1}H, g_2^{-1}H \cdots g_k^{-1}H$ are all the distinct left cosets of H in G, i.e.,
$$G = g_1^{-1}H \cup g_2^{-1}H \cup \ldots \cup g_k^{-1}H$$

88. Prove that if H is a subgroup of a group G, $g \in G$ and if $x \in Hg$, then $Hg = Hx$.

4.7 Normal Subgroups and Quotient Groups

We have seen an example of a group G and a subgroup H of G for which there is an element $g \in G$ and
$$gH \neq Hg$$
For this example see Exercise 76, where $G = S_3$, $H = \{I, (12)\}$ and $g = (13)$. The following results show that a certain class of subgroups, called *normal* subgroups, have the property that $gH = Hg$ for all $g \in G$.

DEFINITION. A subgroup H of a group G is said to be a *normal* subgroup of G if and only if for each $g \in G$
$$ghg^{-1} \in H \text{ for all } h \in H$$

Lemma. If H is a normal subgroup of a group G, then $gH = Hg$ for all $g \in G$.

Proof. We show that (1) $gH \subseteq Hg$ and (2) $Hg \subseteq gH$.
(1) If $x \in gH$, then $x = gh$ for some $h \in H$. Then $xg^{-1} = ghg^{-1} \in H$ since H is normal in G, i.e., $xg^{-1} = h_1$ for some $h_1 \in H$. Therefore, from $xg^{-1} = h_1$ we have $x = h_1g \in Hg$. Thus $gH \subseteq Hg$.
(2) If $y \in Hg$, then $y = hg$ for some $h \in H$. Then $g^{-1}y = g^{-1}hg = g^{-1}h(g^{-1})^{-1} \in H$ since H is normal. Hence, $g^{-1}y = h_2$ for some $h_2 \in H$; i.e., $y = gh_2 \in gH$. Thus $Hg \subseteq gH$.

Lemma. If H is a subgroup of group G and if $gH = Hg$ for all $g \in G$, then H is a normal subgroup of G.

Proof. We need to show that $ghg^{-1} \in H$ for each $g \in G$ and all $h \in H$. It is given that $gH = Hg$ so that there is an $h_1 \in H$ such that $gh = h_1g$. Then $(gh)g^{-1} = (h_1g)g^{-1} = h_1 \in H$. Therefore, H is normal in G.

EXAMPLE 4-19: In $S_3 = \{I, (12), (13), (23), (123), (132)\}$, consider the left and right cosets of the subgroup $A_3 = \{I, (123), (132)\} = H$. Since H has order 3, there can be only two distinct left cosets of H and only two distinct right cosets of H.

Right Cosets of H:
$H, H(12) = \{(12), (23), (13)\}$

Left Cosets of H:
$H, (12)H = \{(12), (13), (23)\}$

Thus we see that $H(12) = (12)H$ and since $H(23) = H(12)$ and $(23)H = (12)H$, $H(23) = (23)H$. Also $(13)H = H(13)$. We conclude that A_3 is a normal subgroup of S_3.

EXAMPLE 4-20: By Exercise 76, the subgroup $H = \{I, (12)\}$ of S_3 is not normal in S_3 since $(13)H \neq H(13)$.

Lemma. Let H be a subgroup of group G and suppose that $[G:H] = 2$. Then H is a normal subgroup of G.

Proof. Since $[G:H] = 2$, there are 2 left cosets of H in G and also 2 right cosets of H in G. H is both a right and a left coset of H in G and if $g \in G$, $g \notin H$, then gH, Hg are left and right cosets of G respectively. Therefore, H, Hg are the two right cosets of H in G and H, gH are the two left cosets of H in G. Hence,

$$G = H \cup Hg = H \cup gH$$

It then follows that $G \cap H = Hg$ and $G \cap H = gH$ so that $gH = Hg$. We then conclude that $gH = Hg$ for all $g \in G$ and that H is a normal subgroup of G.

EXAMPLE 4-21: The symmetric group S_n has order $n!$ and the subgroup A_n of S_n has order $n!/2$. Therefore, $[S_n : A_n] = 2$. It follows that A_n is a normal subgroup of S_n.

One of the important reasons for studying normal subgroups of a group is that the cosets (left or right) of a normal subgroup are a group with respect to an appropriate binary operation. If H is a normal subgroup of a group G, we associate with two cosets Hg_1 and Hg_2 the coset $Hg_1 \cdot g_2$, i.e.,

$$(Hg_1, Hg_2) \longrightarrow Hg_1 \cdot g_2 \quad \text{for all } g_1, g_2 \in G$$

This is also expressed as

$$Hg_1 \cdot Hg_2 = Hg_1 \cdot g_2$$

Notice that the fact that H is normal was not used in making this definition; it could have been made for *any* subgroup H. However, we wish to show that $(Hg_1, Hg_2) \to Hg_1 \cdot g_2$ is a mapping, and in the absence of the normality of H, this cannot be done. To show that this is a mapping we must show that if $Hg_1 = Hg_3$ and $Hg_2 = Hg_4$, then $Hg_1 \cdot g_2 = Hg_3 \cdot g_4$. We have

$$\begin{aligned}
Hg_1 \cdot g_2 &= (Hg_1) \cdot g_2 \\
&= (Hg_3) \cdot g_2 && \text{substituting } Hg_3 \text{ for } Hg_1 \\
&= (g_3 H) \cdot g_2 && \text{using the fact that } H \text{ is normal} \\
&= g_3 (H \cdot g_2) \\
&= g_3 (Hg_4) && \text{substituting } Hg_4 \text{ for } Hg_2 \\
&= (g_3 H) g_4 \\
&= Hg_3 \cdot g_4 && \text{using the fact that } H \text{ is normal}
\end{aligned}$$

Theorem 37 (*Galois*). The set of all right (or left) cosets of a normal subgroup H of a group G is a group with respect to the binary operation defined above.

Proof. It was just shown that a binary operation is defined by
$$g_1 H \cdot g_2 H = g_1 \cdot g_2 H$$

(a) *The identity element is $H = eH$.* For all $g \in G$, $gH \cdot eH = g \cdot eH = gH$ and $eH \cdot gH = e \cdot gH = gH$ so that $eH = H$ is the identity element for the set of right cosets of H.

(b) *The binary operation is associative.* For all $g_1, g_2, g_3 \in G$
$$g_1 H(g_2 H \cdot g_3 H) = g_1 H(g_2 \cdot g_3 H) = g_1 \cdot g_2 \cdot g_3 H$$
$$(g_1 H \cdot g_2 H) \cdot g_3 H = (g_1 \cdot g_2 H) g_3 H = g_1 \cdot g_2 \cdot g_3 H$$

Therefore, $g_1 H(g_2 H \cdot g_3 H) = (g_1 H \cdot g_2 H) g_3 H$.

(c) *Each element gH has an inverse.*
$$gH \cdot g^{-1} H = g \cdot g^{-1} H = eH = H$$
and
$$g^{-1} H \cdot gH = g^{-1} \cdot gH = eH = H$$

Thus $g^{-1} H$ is the inverse of gH.

EXAMPLE 4-22: We showed above that A_3 is a normal subgroup of the group S_3. The left cosets are A_3 and $(12)A_3$. The group table for these cosets is constructed below:

	A_3	$(12)A_3$
A_3	A_3	$(12)A_3$
$(12)A_3$	$(12)A_3$	A_3

EXAMPLE 4-23: In a commutative group G, every subgroup H is normal since $ghg^{-1} = hgg^{-1} = h \cdot e = h \in H$ for all $h \in H$ and $g \in G$. All the groups $(Z_m, \oplus)$ and $(Z_p, \otimes)$ are commutative since
$$\bar{a} \oplus \bar{b} = \overline{a + b} = \overline{b + a} = \bar{b} \oplus \bar{a}$$
and
$$\bar{a} \otimes \bar{b} = \overline{a \cdot b} = \overline{b \cdot a} = \bar{b} \otimes \bar{a}$$

For example, consider $(Z_{12}, \oplus)$. The subgroup $H = [\bar{4}]$ is a normal cyclic subgroup of order 3. We have
$$H\bar{0} = [\bar{4}] = \{\bar{0}, \bar{4}, \bar{8}\}$$
$$H\bar{1} = [\bar{4}]\bar{1} = \{\bar{1}, \bar{5}, \bar{9}\}$$
$$H\bar{2} = [\bar{4}]\bar{2} = \{\bar{2}, \bar{6}, \overline{10}\}$$
$$H\bar{3} = [\bar{4}]\bar{3} = \{\bar{3}, \bar{7}, \overline{11}\}$$

Groups 197

The group table for the cosets is as follows:

	$H\bar{0}$	$H\bar{1}$	$H\bar{2}$	$H\bar{3}$
$H\bar{0}$	$H\bar{0}$	$H\bar{1}$	$H\bar{2}$	$H\bar{3}$
$H\bar{1}$	$H\bar{1}$	$H\bar{2}$	$H\bar{3}$	$H\bar{0}$
$H\bar{2}$	$H\bar{2}$	$H\bar{3}$	$H\bar{0}$	$H\bar{1}$
$H\bar{3}$	$H\bar{3}$	$H\bar{0}$	$H\bar{1}$	$H\bar{2}$

DEFINITION. If H is a normal subgroup of a group G, then the group of all left cosets is called the *quotient group of G by H* and is denoted by G/H.

Observe that for a normal subgroup H of a group G, the order of the quotient group G/H is the index of H in G.

Exercises

	$\bar{1}$	$\bar{3}$	$\bar{5}$	$\bar{9}$	$\bar{11}$	$\bar{13}$
$\bar{1}$	$\bar{1}$	$\bar{3}$	$\bar{5}$	$\bar{9}$	$\bar{11}$	$\bar{13}$
$\bar{3}$	$\bar{3}$	$\bar{9}$	$\bar{1}$	$\bar{13}$	$\bar{5}$	$\bar{11}$
$\bar{5}$	$\bar{5}$	$\bar{1}$	$\bar{11}$	$\bar{3}$	$\bar{13}$	$\bar{9}$
$\bar{9}$	$\bar{9}$	$\bar{13}$	$\bar{3}$	$\bar{11}$	$\bar{1}$	$\bar{5}$
$\bar{11}$	$\bar{11}$	$\bar{5}$	$\bar{13}$	$\bar{1}$	$\bar{9}$	$\bar{3}$
$\bar{13}$	$\bar{13}$	$\bar{11}$	$\bar{9}$	$\bar{5}$	$\bar{3}$	$\bar{1}$

$\{\bar{1}, \bar{3}, \bar{5}, \bar{9}, \bar{11}, \bar{13}\}$

$\{\bar{1}\}$, $\{\bar{1}, \bar{3}, \bar{9}, \bar{13}, \bar{11}, \bar{5}\}$

$\{\bar{1}, \bar{5}, \bar{11}, \bar{13}, \bar{9}, \bar{3}\}$, $\{\bar{1}, \bar{9}, \bar{11}\}$

$\{\bar{1}, \bar{11}, \bar{9}\}$, $\{\bar{1}, \bar{13}\}$

Yes

89. Construct the group table below for $(P_{14}, \otimes)$.

	$\bar{1}$	$\bar{3}$	$\bar{5}$	$\bar{9}$	$\bar{11}$	$\bar{13}$
$\bar{1}$						
$\bar{3}$						
$\bar{5}$						
$\bar{9}$						
$\bar{11}$						
$\bar{13}$						

(a) $P_{14} = $ _____

(b) $[\bar{1}] = $ _____, $[\bar{3}] = $ _____

$[\bar{5}] = $ _____, $[\bar{9}] = $ _____

$[\bar{11}] = $ _____, $[\bar{13}] = $ _____

(c) Is P_{14} a cyclic group? _____

198   Groups

$\bar{3}, \bar{5}$

6

1

1

2

2

3

Yes

Yes

2

	H	H$\bar{3}$
H	H	H$\bar{3}$
H$\bar{3}$	H$\bar{3}$	H

3

[13], [13]$\bar{3}$, [13]$\bar{5}$

	H	H$\bar{3}$	H$\bar{5}$
H	H	H$\bar{3}$	H$\bar{5}$
H$\bar{3}$	H$\bar{3}$	H$\bar{5}$	H
H$\bar{5}$	H$\bar{5}$	H	H$\bar{3}$

Which elements of P_{14} will alone generate P_{14}?

_____

(d) The index of $[\bar{1}]$ in P_{14} is _____

(e) The index of $[\bar{3}]$ in P_{14} is _____

(f) $[P_{14}: [\bar{5}]] = $ _____

(g) $[P_{14}: [\bar{9}]] = $ _____

(h) $[P_{14}: [\overline{11}]] = $ _____

(i) $[P_{14}: [\overline{13}]] = $ _____

(j) Is P_{14} commutative? _____

Is every subgroup of P_{14} normal in P_{14}? _____

(k) What is the order of $P_{14}/[\bar{9}]$? _____

If $H = [\bar{9}]$, make a group table for P_{14}/H below.

	H	H$\bar{3}$
H		
H$\bar{3}$		

(l) What is the order of $P_{14}/[\overline{13}]$? _____

(m) What are the elements of $P_{14}/[\overline{13}]$?

_____

(n) If $H = [\overline{13}]$, complete the table below for $P_{14}/[\overline{13}]$.

	H	H$\bar{3}$	H$\bar{5}$
H			
H$\bar{3}$			
H$\bar{5}$			

Groups 199

⊗	$\bar{1}$	$\bar{2}$	$\bar{4}$	$\bar{7}$	$\bar{8}$	$\overline{11}$	$\overline{13}$	$\overline{14}$
$\bar{1}$	$\bar{1}$	$\bar{2}$	$\bar{4}$	$\bar{7}$	$\bar{8}$	$\overline{11}$	$\overline{13}$	$\overline{14}$
$\bar{2}$	$\bar{2}$	$\bar{4}$	$\bar{8}$	$\overline{14}$	$\bar{1}$	$\bar{7}$	$\overline{11}$	$\overline{13}$
$\bar{4}$	$\bar{4}$	$\bar{8}$	$\bar{1}$	$\overline{13}$	$\bar{2}$	$\overline{14}$	$\bar{7}$	$\overline{11}$
$\bar{7}$	$\bar{7}$	$\overline{14}$	$\overline{13}$	$\bar{4}$	$\overline{11}$	$\bar{2}$	$\bar{1}$	$\bar{8}$
$\bar{8}$	$\bar{8}$	$\bar{1}$	$\bar{2}$	$\overline{11}$	$\bar{4}$	$\overline{13}$	$\overline{14}$	$\bar{7}$
$\overline{11}$	$\overline{11}$	$\bar{7}$	$\overline{14}$	$\bar{2}$	$\overline{13}$	$\bar{1}$	$\bar{8}$	$\bar{4}$
$\overline{13}$	$\overline{13}$	$\overline{11}$	$\bar{7}$	$\bar{1}$	$\overline{14}$	$\bar{8}$	$\bar{4}$	$\bar{2}$
$\overline{14}$	$\overline{14}$	$\overline{13}$	$\overline{11}$	$\bar{8}$	$\bar{7}$	$\bar{4}$	$\bar{2}$	$\bar{1}$

$\{\bar{1}\}$, $\{\bar{1}, \bar{2}, \bar{4}, \bar{8}\}$

$\{\bar{1}, \bar{4}\}$, $\{\bar{1}, \bar{7}, \bar{4}, \overline{13}\}$

$\{\bar{1}, \bar{8}, \bar{4}, \bar{2}\}$, $\{\bar{1}, \overline{11}\}$

$\{\bar{1}, \overline{13}, \bar{4}, \bar{7}\}$, $\{\bar{1}, \overline{14}\}$

No

Because $[\bar{x}] \neq P_{15}$ for all $\bar{x} \in P_{15}$

90. Construct the group table below for $(P_{15}, \otimes)$.

⊗	$\bar{1}$	$\bar{2}$	$\bar{4}$	$\bar{7}$	$\bar{8}$	$\overline{11}$	$\overline{13}$	$\overline{14}$
$\bar{1}$								
$\bar{2}$								
$\bar{4}$								
$\bar{7}$								
$\bar{8}$								
$\overline{11}$								
$\overline{13}$								
$\overline{14}$								

(a) $P_{15} = \{\bar{1}, \bar{2}, \bar{4}, \bar{7}, \bar{8}, \overline{11}, \overline{13}, \overline{14}\}$

(b) $[\bar{1}] =$ _____, $[\bar{2}] =$ _____

$[\bar{4}] =$ _____, $[\bar{7}] =$ _____

$[\bar{8}] =$ _____, $[\overline{11}] =$ _____

$[\overline{13}] =$ _____, $[\overline{14}] =$ _____

(c) Is P_{15} a cyclic group? _____

Why? _____

Groups

	$\bar{1}$	$\bar{2}$	$\bar{7}$	$\bar{4}$	$\overline{14}$	$\bar{8}$	$\overline{13}$	$\overline{11}$
$\bar{1}$	$\bar{1}$	$\bar{2}$	$\bar{7}$	$\bar{4}$	$\overline{14}$	$\bar{8}$	$\overline{13}$	$\overline{11}$
$\bar{2}$	$\bar{2}$	$\bar{4}$	$\overline{14}$	$\bar{8}$	$\overline{13}$	$\bar{1}$	$\overline{11}$	$\bar{7}$
$\bar{7}$	$\bar{7}$	$\overline{14}$	$\bar{4}$	$\overline{13}$	$\bar{8}$	$\overline{11}$	$\bar{1}$	$\bar{2}$
$\bar{4}$	$\bar{4}$	$\bar{8}$	$\overline{13}$	$\bar{1}$	$\overline{11}$	$\bar{2}$	$\bar{7}$	$\overline{14}$
$\overline{14}$	$\overline{14}$	$\overline{13}$	$\bar{8}$	$\overline{11}$	$\bar{1}$	$\bar{7}$	$\bar{2}$	$\bar{4}$
$\bar{8}$	$\bar{8}$	$\bar{1}$	$\overline{11}$	$\bar{2}$	$\bar{7}$	$\bar{4}$	$\overline{14}$	$\overline{13}$
$\overline{13}$	$\overline{13}$	$\overline{11}$	$\bar{1}$	$\bar{7}$	$\bar{2}$	$\overline{14}$	$\bar{4}$	$\bar{8}$
$\overline{11}$	$\overline{11}$	$\bar{7}$	$\bar{2}$	$\overline{14}$	$\bar{4}$	$\overline{13}$	$\bar{8}$	$\bar{1}$

P_{15}

	$\bar{1}$	$\bar{4}$	$\overline{11}$	$\overline{14}$
$\bar{1}$	$\bar{1}$	$\bar{4}$	$\overline{11}$	$\overline{14}$
$\bar{4}$	$\bar{4}$	$\bar{1}$	$\overline{14}$	$\overline{11}$
$\overline{11}$	$\overline{11}$	$\overline{14}$	$\bar{1}$	$\bar{4}$
$\overline{14}$	$\overline{14}$	$\overline{11}$	$\bar{4}$	$\bar{1}$

$\{\bar{1}, \bar{4}, \overline{11}, \overline{14}\}$

8

2

4

2

2

4

1

(d) Compute $[\bar{2}, \bar{7}]$ in the table below:

	$\bar{1}$	$\bar{2}$	$\bar{7}$	$\bar{4}$	$\overline{14}$	$\bar{8}$	$\overline{13}$	$\overline{11}$
$\bar{1}$								
$\bar{2}$								
$\bar{7}$								
$\bar{4}$								
$\overline{14}$								
$\bar{8}$								
$\overline{13}$								
$\overline{11}$								

Therefore, $[\bar{2}, \bar{7}] = $ _____

(e) Compute $[\bar{4}, \overline{11}, \overline{14}]$ in the table below:

	$\bar{1}$	$\bar{4}$	$\overline{11}$	$\overline{14}$
$\bar{1}$				
$\bar{4}$				
$\overline{11}$				
$\overline{14}$				

Therefore, $[\bar{4}, \overline{11}, \overline{14}] = $ _____

(f) $[P_{15} : [\bar{1}]] = $ _____

$[P_{15} : [\bar{2}]] = $ _____

$[P_{15} : [\bar{4}]] = $ _____

$[P_{15} : [\bar{7}]] = $ _____

$[P_{15} : [\bar{4}]] = $ _____

$[P_{15} : [\overline{11}]] = $ _____

$[P_{15} : [\bar{2}, \bar{7}]] = $ _____

2

$H\bar{1}, H\bar{2}$

$H\bar{1}, H\bar{2}, H\bar{4}, H\bar{8}$

$H\bar{1}, H\bar{2}, H\bar{4}, H\bar{7}$
Yes
Yes
$H\bar{1}, H\bar{2}$

	$H\bar{1}$	$H\bar{2}$
$H\bar{1}$	$H\bar{1}$	$H\bar{2}$
$H\bar{2}$	$H\bar{2}$	$H\bar{1}$

$H\bar{1}, H\bar{2}, H\bar{4}, H\bar{8}$

	$H\bar{1}$	$H\bar{2}$	$H\bar{4}$	$H\bar{8}$
$H\bar{1}$	$H\bar{1}$	$H\bar{2}$	$H\bar{4}$	$H\bar{8}$
$H\bar{2}$	$H\bar{2}$	$H\bar{4}$	$H\bar{8}$	$H\bar{1}$
$H\bar{4}$	$H\bar{4}$	$H\bar{8}$	$H\bar{1}$	$H\bar{2}$
$H\bar{8}$	$H\bar{8}$	$H\bar{1}$	$H\bar{2}$	$H\bar{4}$

$[P_{15} : [\bar{4}, \bar{11}, \bar{14}]] = $ _____

(g) Let $H = [\bar{7}]$. Compute the right cosets of H in P_{15}. _____

(h) Let $H = [\bar{11}]$. Compute the right cosets of H in P_{15}. _____

(i) Let $H = [\bar{14}]$. Compute the right cosets of H in P_{15}. _____

(j) Is $(P_{15}, \otimes)$ a commutative group? _____

(k) Is each subgroup of $(P_{15}, \otimes)$ normal? _____

(l) The elements of $P_{15}/[\bar{7}]$ are _____

Make a group table below for $P_{15}/[\bar{7}]$.

	$H\bar{1}$	$H\bar{2}$
$H\bar{1}$		
$H\bar{2}$		

(m) The elements of $P_{15}/[\bar{11}]$ are _____

Make a group table below for $P_{15}/[\bar{11}]$.

	$H\bar{1}$	$H\bar{2}$	$H\bar{4}$	$H\bar{8}$
$H\bar{1}$				
$H\bar{2}$				
$H\bar{4}$				
$H\bar{8}$				

$H\bar{1}, H\bar{2}, H\bar{4}, H\bar{7}$

	$H\bar{1}$	$H\bar{2}$	$H\bar{4}$	$H\bar{7}$
$H\bar{1}$	$H\bar{1}$	$H\bar{2}$	$H\bar{4}$	$H\bar{7}$
$H\bar{2}$	$H\bar{2}$	$H\bar{4}$	$H\bar{7}$	$H\bar{1}$
$H\bar{4}$	$H\bar{4}$	$H\bar{7}$	$H\bar{1}$	$H\bar{2}$
$H\bar{7}$	$H\bar{7}$	$H\bar{1}$	$H\bar{2}$	$H\bar{4}$

$H\bar{1}, H\bar{2}, H\bar{4}, H\bar{7}$

	$H\bar{1}$	$H\bar{2}$	$H\bar{4}$	$H\bar{7}$
$H\bar{1}$	$H\bar{1}$	$H\bar{2}$	$H\bar{4}$	$H\bar{7}$
$H\bar{2}$	$H\bar{2}$	$H\bar{4}$	$H\bar{7}$	$H\bar{1}$
$H\bar{4}$	$H\bar{4}$	$H\bar{7}$	$H\bar{1}$	$H\bar{2}$
$H\bar{7}$	$H\bar{7}$	$H\bar{1}$	$H\bar{2}$	$H\bar{4}$

12

$(ab)(ac)$

even

$(123), (132), (124), (142), (134), (143)$

$(243), (234)$

(n) The elements of $P_{15}/[\bar{1}]]$ are _____

Make a group table below for $P_{15}/[\overline{14}]$.

	$H\bar{1}$	$H\bar{2}$	$H\bar{4}$	$H\bar{7}$
$H\bar{1}$				
$H\bar{2}$				
$H\bar{4}$				
$H\bar{7}$				

(o) The elements of $P_{15}/[\overline{14}]$ are _____

Make a group table below for $P_{15}/[\overline{14}]$.

	$H\bar{1}$	$H\bar{2}$	$H\bar{4}$	$H\bar{7}$
$H\bar{1}$				
$H\bar{2}$				
$H\bar{4}$				
$H\bar{7}$				

91. (a) The order of A_4 is _____

(b) Every 3-cycle (abc) may be written as a product of transpositions as follows: _____

Therefore, every 3-cycle is an _____ permutation and every 3-cycle of S_4 belongs to A_4.

(c) List all the 3-cycles of S_4 which involve 1: _____

(d) List all the 3-cycles of S_4 which involve 2 and are not in (c) above: _____

(e) The elements listed in (c) and (d) above are all the 3-cycles that are members of A_4. Together with the identity, these give the nine elements I, (123), (132), (124), (142), (134), (143), (243), (234). The subgroup of A_4 generated by these eight 3-cycles must have order at least 9. Since A_4 has order 12, the only possible orders for subgroups are 1, 2, 3, 4, 6, 12. Therefore, the subgroup generated by these eight 3-cycles must be of order 12; i.e., the subgroup generated by these 3-cycles is A_4. Compute A_4 from these 3-cycles by the table method, on p. 204.

(f) Compute [(12)(34), (13)(24)]

	I	(12)(34)	(13)(24)	(14)(23)
I				
(12)(34)				
(13)(24)				
(14)(23)				

Therefore, [(12)(34), (13)(24)] = _____

(g) If H = [(12)(34), (13)(24)], then the order of H is _____ and $[A_4 : H]$ = _____

Compute the right cosets of H: _____

where HI = _____

$H(123)$ = _____

$H(124)$ = _____

Compute the left cosets of H:

_____, where

	I	(12)(34)	(13)(24)	(14)(23)
I	I	(12)(34)	(13)(24)	(14)(23)
(12)(34)	(12)(34)	I	(14)(23)	(13)(24)
(13)(24)	(13)(24)	(14)(23)	I	(12)(34)
(14)(23)	(14)(23)	(13)(24)	(12)(34)	I

{I, (12)(34), (13)(24), (14)(23)}

4, 3

HI, $H(123)$, $H(124)$

{I, (12)(34), (13)(24), (14)(23)}

{(123), (134), (243), (142)}

{(124), (143), (132), (234)}

IH, (123)H, (124)H

204 Groups

	I	(123)	(132)	(124)	(142)	(134)	(143)	(234)	(243)	(14)(23)	(12)(34)	(13)(24)
I												
(123)												
(132)												
(124)												
(142)												
(134)												
(143)												
(234)												
(243)												
(14)(23)												
(12)(34)												
(13)(24)												

	I	(123)	(132)	(124)	(142)	(134)	(143)	(234)	(243)	(14)(23)	(12)(34)	(13)(24)
I	I	(123)	(132)	(124)	(142)	(134)	(143)	(234)	(243)	(14)(23)	(12)(34)	(13)(24)
(123)	(123)	(132)	I	(14)(23)	(234)	(124)	(12)(34)	(13)(24)	(143)	(134)	(243)	(142)
(132)	(132)	I	(123)	(134)	(13)(24)	(14)(23)	(243)	(142)	(12)(34)	(124)	(143)	(234)
(124)	(124)	(13)(24)	(243)	(142)	I	(12)(34)	(123)	(134)	(14)(23)	(132)	(234)	(143)
(142)	(142)	(143)	(14)(23)	I	(124)	(234)	(13)(24)	(12)(34)	(132)	(243)	(134)	(123)
(134)	(134)	(234)	(12)(34)	(13)(24)	(132)	(143)	I	(14)(23)	(124)	(123)	(142)	(243)
(143)	(143)	(14)(23)	(142)	(243)	(12)(34)	I	(134)	(123)	(13)(24)	(234)	(132)	(124)
(234)	(234)	(12)(34)	(134)	(123)	(14)(23)	(13)(24)	(142)	(243)	I	(143)	(124)	(132)
(243)	(243)	(124)	(13)(24)	(12)(34)	(143)	(132)	(14)(23)	I	(234)	(142)	(123)	(134)
(14)(23)	(14)(23)	(142)	(143)	(234)	(123)	(243)	(132)	(124)	(134)	I	(13)(24)	(12)(34)
(12)(34)	(12)(34)	(134)	(234)	(143)	(243)	(123)	(124)	(132)	(142)	(13)(24)	I	(14)(23)
(13)(24)	(13)(24)	(243)	(124)	(132)	(134)	(142)	(234)	(143)	(123)	(12)(34)	(14)(23)	I

{I, (12)(34), (13)(24), (14)(23)}

{(123), (243), (142), (134)}

{(124), (234), (143), (132)}

Yes

H is a normal subgroup of A_4.

$IH =$ _____

$(123)H =$ _____

$(124)H =$ _____

(h) Is it true that $H(123) = (123)H$ and $H(124) = (124)H$? _____

What do you conclude from this? _____

92. Determine all the normal subgroups of the groups A_1, A_2, A_3.

93. Let G be a group and let $g \in G$. Define $K = \{gyg^{-1} | y \in G\}$. Prove that $K = G$.

94. Let G be a group and define gRh, for $g, h \in G$, to mean that there is an element $x \in G$ such that $g = xhx^{-1}$. (If gRh, g, h are said to be *conjugate* elements).
    (a) Show that R is an equivalence relation on G.
    (b) The equivalence classes associated with R are called the *conjugate* classes of G. Determine the conjugate classes of A_4.
    (c) If H is a normal subgroup of G, show that H is a conjugate class of G.

4.8 Isomorphism

It may have occurred to the reader by now that groups sometimes differ in name only. For example, the two groups $(Z_4, \oplus)$ and $(Z_5^*, \otimes)$ are in most respects very much alike. It is obvious that they are alike in "size" since both have order 4. It may not be so obvious that the binary operations are similar. The similarity of the binary operations is indicated to some extent by the fact that each is a cyclic group:

Z_4

$\oplus$	$\bar{0}$	$\bar{1}$	$\bar{2}$	$\bar{3}$
$\bar{0}$	$\bar{0}$	$\bar{1}$	$\bar{2}$	$\bar{3}$
$\bar{1}$	$\bar{1}$	$\bar{2}$	$\bar{3}$	$\bar{0}$
$\bar{2}$	$\bar{2}$	$\bar{3}$	$\bar{0}$	$\bar{1}$
$\bar{3}$	$\bar{3}$	$\bar{0}$	$\bar{1}$	$\bar{2}$

Z_5^*

$\otimes$	$\bar{1}$	$\bar{2}$	$\bar{3}$	$\bar{4}$
$\bar{1}$	$\bar{1}$	$\bar{2}$	$\bar{3}$	$\bar{4}$
$\bar{2}$	$\bar{2}$	$\bar{4}$	$\bar{1}$	$\bar{3}$
$\bar{3}$	$\bar{3}$	$\bar{1}$	$\bar{4}$	$\bar{2}$
$\bar{4}$	$\bar{4}$	$\bar{3}$	$\bar{2}$	$\bar{1}$

$Z_4 = [\bar{1}]$ and $Z_5^* = [\bar{2}]$. Thus each set consists of powers of a single element. The likeness between the binary operations can be further emphasized by introducing a mapping of Z_4 onto Z_5^* as follows:

$$T = \begin{pmatrix} \bar{0} & \bar{1} & \bar{2} & \bar{3} \\ \bar{1} & \bar{2} & \bar{4} & \bar{3} \end{pmatrix} \begin{matrix} \longleftarrow \text{elements of } Z_4 \\ \longleftarrow \text{elements of } Z_5^* \end{matrix}$$

Consider now the element $\bar{1} \oplus \bar{2} = \bar{3}$. The mapping above maps $\bar{1}$ into $\bar{2}$ and $\bar{2}$ into $\bar{4}$. The images of all the elements involved may be pictured below:

$$\begin{array}{ccccc} \bar{1} & \oplus & \bar{2} & = & \bar{3} \\ T\downarrow & & T\downarrow & & \uparrow \\ \bar{2} & \otimes & \bar{4} & = & \bar{3} \end{array} \quad \text{This agrees with mapping } T.$$

Observe that if the arrow is drawn on the right-hand side above, the result agrees with the given mapping. Again, consider $\bar{2} \oplus \bar{3}$:

$$\begin{array}{ccccc} \bar{2} & \oplus & \bar{3} & = & \bar{1} \\ T\downarrow & & T\downarrow & & \uparrow \\ \bar{4} & \otimes & \bar{3} & = & \bar{2} \end{array} \quad \text{This agrees with mapping } T.$$

Now consider the element $\bar{3} \oplus \bar{1}$ in the following diagram:

$$\begin{array}{ccccc} \bar{3} & \oplus & \bar{1} & = & \bar{0} \\ T\downarrow & & T\downarrow & & T\downarrow \\ \Box & \otimes & \Box & = & \Box \end{array}$$

Without referring to the group table for $(Z_5^*, \otimes)$, fill in the boxes by using only the mapping T:

$$\begin{array}{ccccc} \bar{3} & \oplus & \bar{1} & = & \bar{0} \\ T\downarrow & & T\downarrow & & T\downarrow \\ \boxed{\bar{3}} & \otimes & \boxed{\bar{2}} & = & \boxed{\bar{1}} \end{array}$$

The result is the statement $\bar{3} \otimes \bar{2} = \bar{1}$, which agrees with the group table for $(Z_5^*, \otimes)$. If $\bar{a}, \bar{b}$ are any members of Z_4, and if we use the mapping T to fill in the blocks below, we obtain as a result a statement $\bar{x} \otimes \bar{y} = \bar{z}$.

$$\begin{array}{ccc} \bar{a} & \oplus \ \bar{b} & = \ \bar{c} \\ T\downarrow & T\downarrow & T\downarrow \\ \boxed{\bar{x}} & \otimes \ \boxed{\bar{y}} & = \boxed{\bar{z}} \end{array}$$

In this case—no matter which elements $\bar{a}$, $\bar{b}$ we use—the resulting statement obtained by using only the group table for $(Z_4, \oplus)$ and the mapping T is always a true statement. It is therefore possible to construct the group table for $(Z_5^*, \otimes)$ by using only the group table for $(Z_4, \oplus)$ and the mapping T. It is in this sense that the two binary operations $\oplus$ and $\otimes$ are alike.

If we abstract the essential features from the above situation, we obtain the following description. There are two groups (G_1, B_1), (G_2, B_2) with binary operations B_1, B_2 respectively and a 1-1 mapping T of G_1 onto G_2 such that the following is true: if for $a, b \in G_1$

$$T: a \longrightarrow x, x \in G_2$$
$$T: b \longrightarrow y, y \in G_2$$

and if $aB_1 b = c$, $xB_2 y = z$, then

$$T: c \longrightarrow z$$

Written in an alternate notation we have $T(c) = z$ or, since $c = aB_1 b$ and $z = xB_2 y$,

$$T(aB_1 b) = xB_2 y$$

Finally, since $T(a) = x$ and $T(b) = y$

$$T(aB_1 b) = T(a)B_2 T(b) \qquad (4\text{-}1)$$

Normally, when the binary operations B_1, B_2 are not specified and when no confusion results, the notation $\cdot$ is used for both B_1 and B_2. Then (4-1) is

$$T(a \cdot b) = T(a) \cdot T(b) \qquad (4\text{-}2)$$

DEFINITION. Let G_1, G_2 be groups. G_1 is *isomorphic* to G_2 if and only if there is a 1-1 mapping T of G_1 onto G_2 such that

$$T(a \cdot b) = T(a) \cdot T(b) \quad \text{for all } a, b \in G_1$$

The mapping T is called an *isomorphism*.

EXAMPLE 4-24: The group $(Z_4, \oplus)$ is isomorphic to the group $(Z_5^*, \otimes)$ with respect to the mapping T given above.

Theorem 38. If G_1 and G_2 are both cyclic groups of finite order n, then G_1 is isomorphic to G_2.

Proof. Since G_1 and G_2 are cyclic groups, there are elements $g_1 \in G_1, g_2 \in G_2$ such that $G_1 = [g_1]$ and $G_2 = [g_2]$. Both G_1 and G_2 have order n so that

$$G_1 = [g_1] = \{e_1, g_1, g_1^2, \ldots, g_1^{n-1}\}$$
$$G_2 = [g_2] = \{e_2, g_2, g_2^2, \ldots, g_2^{n-1}\}$$

where $g_1^n = e_1$ (the identity of G_1) and $g_2^n = e_2$ (the identity of G_2). Define a mapping T as follows:

$$T: \quad g_1^t \longrightarrow g_2^t, \quad 0 \leq t < n$$

We now show that T is a 1-1 mapping of G_1 onto G_2. Suppose $g_1^t = g_1^s$, where t, s are integers such that $0 \leq t < n$ and $0 \leq s < n$. Then if $t \geq s$, $t - s \geq 0$ and from $g_1^t = g_1^s$

$$g_1^{t-s} = e_1$$

Since $t - s < n$ and n is the order of g_1, we conclude that $t - s = 0$ or $t = s$. Hence, $g_2^t = g_2^s$. Therefore, T is a mapping. The argument to show that T is 1-1 is similar. T is clearly an onto mapping. Next, we have

$$T(g_1^t \cdot g_1^s) = T(g_1^{t+s}) = g_2^{t+s} = g_2^t \cdot g_2^s = T(g_1^t) \cdot T(g_1^s)$$

Therefore, G_1 is isomorphic to G_2.

Theorem 38 implies that there is essentially only one cyclic group of order n. We thus speak of *the* cyclic group of order n and denote this group by $C(n)$. Since all groups of prime order are cyclic, we may conclude that there is only one group of a given prime order p, namely $C(p)$. For example, there is only one group of order 2, 3, 5, 7, etc.

Theorem 39. If G_1 and G_2 are both infinite cyclic groups, then G_1 is isomorphic to G_2.

Proof. We know that $G_1 = [g_1]$, $G_2 = [g_2]$ for some $g_1 \in G_1$, $g_2 \in G_2$. We also know, since G_1, G_2 are infinite, that $g_1^n \neq e_1$, $g_2^n \neq e_2$ if $n \neq 0$. Then define T as follows:

$$T: \quad g_1^t \longrightarrow g_2^t, \quad \text{for all } t \in Z$$

T is a mapping, since if $g_1^t = g_1^s$, then

$$g_1^{t-s} = e_1$$

This implies that $t - s = 0$ or $t = s$. Then $g_2^t = g_2^s$ and T is a mapping. In the same way, T is shown to be a 1-1 mapping. Also, T is clearly onto. To complete the proof

$$T(g_1^t \cdot g_1^s) = T(g_1^{t+s}) = g_2^{t+s} = g_2^t \cdot g_2^s = T(g_1^t) \cdot T(g_1^s)$$

Therefore, G_1 is isomorphic to G_2.

Theorem 39 indicates that there is essentially only one infinite cyclic group. This is denoted by $C(\infty)$. The group $(Z, +)$ is an infinite cyclic group: $Z = [1]$. Thus every infinite cyclic group is isomorphic to $(Z, +)$.

Theorem 40 (*Cayley*). If G is a group of finite order n, then G is isomorphic to a subgroup of S_n.

Proof. Since G is finite and has order n, we may write $G = \{g_1, g_2, g_3, \ldots, g_n\}$, where $g_1 = e$, the identity of G. We now consider the mappings R_g for each $g \in G$. These may be expressed as follows:

$$R_g = \begin{pmatrix} g_1 & g_2 & g_3 & \cdots & g_n \\ g_1 \cdot g & g_2 \cdot g & g_3 \cdot g & \cdots & g_n \cdot g \end{pmatrix}$$

Or we may write $R_g: g_i \longrightarrow g_i \cdot g$ for each $i \in \{1, 2, 3, \ldots, n\}$. We have already proved that these mappings are 1–1 and onto G (see Theorem 23, p. 151). Since these R_g are 1–1 mappings of a set of n elements onto itself, they are permutations of n elements and so belong to S_n. Now define $G(R) = \{R_g \mid g \in G\}$. We prove next that $G(R)$ is a subgroup of S_n.

(1) *Products of mappings is a binary operation on $G(R)$.* If $R_g \in G(R)$ and $R_h \in G(R)$, then

$$R_g: x \longrightarrow x \cdot g \quad \text{and} \quad R_h: x \longrightarrow x \cdot h$$

Hence,

$$R_g \cdot R_h: x \longrightarrow x \cdot g \longrightarrow (x \cdot g)h = x \cdot (gh)$$

i.e.,

$$R_g \cdot R_h: x \longrightarrow x \cdot (gh)$$

But also

$$R_{g \cdot h}: x \longrightarrow x \cdot (gh)$$

Therefore,

$$R_g \cdot R_h = R_{g \cdot h}$$

and $R_{g \cdot h} \in G(R)$ since $g \cdot h \in G$, which implies that $R_g \cdot R_h \in G(R)$.

(2) *The element $R_e \in G(R)$ is the identity element.* For any $R_g \in G(R)$

$$R_e \cdot R_g = R_{e \cdot g} = R_g \quad \text{and} \quad R_g \cdot R_e = R_{g \cdot e} = R_g$$

(3) *Each element of $G(R)$ has an inverse in $G(R)$.* If $R_g \in G(R)$, then $R_{g^{-1}} \in G(R)$ and

$$R_g \cdot R_{g^{-1}} = R_{g \cdot g^{-1}} = R_e$$
$$R_{g^{-1}} \cdot R_g = R_{g^{-1} \cdot g} = R_e$$

To complete the proof, it only remains to show that G is isomorphic to $G(R)$. Define T as follows:

$$T: g \longrightarrow R_g$$

(4) *T is a mapping.* For if $g = h$, then $R_g = R_h$.

(5) *T is a 1-1 mapping.* If $R_g = R_h$, then from
$$R_g: x \longrightarrow x \cdot g$$
$$R_h: x \longrightarrow x \cdot h$$
it follows that $x \cdot g = x \cdot h$ for all $x \in G$. Hence, $g = h$ and T is 1-1.

(6) *T is a mapping of G onto G(R).* Given any $R_g \in G(R)$, the element $g \in G$ maps onto R_g by definition of T.

(7) *The equation* $T(g \cdot h) = T(g) \cdot T(h)$ *holds for all* $g, h \in G$.
$$T(g \cdot h) = R_{g \cdot h} = R_g \cdot R_h = T(g) \cdot T(h)$$

The proof is now complete, since we have shown that there is a 1-1 mapping T of G onto $G(R)$ such that $T(g \cdot h) = T(g) \cdot T(h)$. Thus by definition G is isomorphic to $G(R)$ and $G(R)$ is a subgroup of S_n.

DEFINITION. If G is a group, $G(R)$ is called the *right regular representation* of G.

The result of Theorem 40 implies that groups of finite order may be studied by investigating the groups S_n and their subgroups. We may also conclude that there are only finitely many groups of a given finite order, since each S_n has only a finite number of subgroups.

EXAMPLE 4-25: Obtain the right regular representation of the group $(Z_5^*, \otimes)$.

Solution: The group table is

x	$\bar{1}$	$\bar{2}$	$\bar{3}$	$\bar{4}$
$\bar{1}$	$\bar{1}$	$\bar{2}$	$\bar{3}$	$\bar{4}$
$\bar{2}$	$\bar{2}$	$\bar{4}$	$\bar{1}$	$\bar{3}$
$\bar{3}$	$\bar{3}$	$\bar{1}$	$\bar{4}$	$\bar{2}$
$\bar{4}$	$\bar{4}$	$\bar{3}$	$\bar{2}$	$\bar{1}$

$$R_{\bar{1}} = \begin{pmatrix} \bar{1} & \bar{2} & \bar{3} & \bar{4} \\ \bar{1} & \bar{2} & \bar{3} & \bar{4} \end{pmatrix} \quad R_{\bar{2}} = \begin{pmatrix} \bar{1} & \bar{2} & \bar{3} & \bar{4} \\ \bar{2} & \bar{4} & \bar{1} & \bar{3} \end{pmatrix}$$

$$R_{\bar{3}} = \begin{pmatrix} \bar{1} & \bar{2} & \bar{3} & \bar{4} \\ \bar{3} & \bar{1} & \bar{4} & \bar{2} \end{pmatrix} \quad R_{\bar{4}} = \begin{pmatrix} \bar{1} & \bar{2} & \bar{3} & \bar{4} \\ \bar{4} & \bar{3} & \bar{2} & \bar{1} \end{pmatrix}$$

Then $G(R) = \{R_1, R_2, R_3, R_4\}$. If we drop the "bars"* and write these in cycle notation, we have
$$G(R) = \{I, (1243), (1342), (14)(23)\}$$

---

*From this point, the bar-notation will no longer be used in the groups $(Z_n, \oplus)$, $(Z_p^*, \otimes)$ and P_n since the meaning will always be clear without this added precision. For example, we will write $Z_5 = \{0, 1, 2, 3, 4\}$ instead of $Z_5 = \{\bar{0}, \bar{1}, \bar{2}, \bar{3}, \bar{4}\}$.

Below we have superimposed the table for $G(R)$ on the table for $(Z_5^*, \otimes)$. Observe that corresponding elements occur in the same blocks.

	$\bar{1}$ I	$\bar{2}$ (1243)	$\bar{3}$ (1342)	$\bar{4}$ $(14)(23)$
$\bar{1}$ I	$\bar{1}$ I	$\bar{2}$ (1243)	$\bar{3}$ (1342)	$\bar{4}$ $(14)(23)$
$\bar{2}$ (1243)	$\bar{2}$ (1243)	$\bar{4}$ $(14)(23)$	$\bar{1}$ I	$\bar{3}$ (1342)
$\bar{3}$ (1342)	$\bar{3}$ (1342)	$\bar{1}$ I	$\bar{4}$ $(14)(23)$	$\bar{2}$ (1243)
$\bar{4}$ $(14)(23)$	$\bar{4}$ $(14)(23)$	$\bar{3}$ (1342)	$\bar{2}$ (1243)	$\bar{1}$ I

Lemma. Let T be an isomorphism from a group G_1 onto a group G_2. Then
(1) If e is the identity of G_1, $T(e)$ is the identity of G_2.
(2) For each $g \in G_1$, $T(g^{-1})$ is the inverse of $T(g)$.

Proof. (1) Suppose $g_2 \in G_2$. T is a mapping from G_1 onto G_2 and, hence, there is an element $g_1 \in G_1$ such that $g_2 = T(g_1)$. Then

$$\begin{aligned} g_2 \cdot T(e) &= T(g_1) \cdot T(e) & &\text{substituting for } g_2 \\ &= T(g_1 \cdot e) & &T \text{ is an isomorphism} \\ &= T(g_1) & &\text{substituting } g_1 \text{ for } g_1 \cdot e \\ &= g_2 & &\text{substituting } g_2 \text{ for } T(g_1) \end{aligned}$$

Therefore, $g_2 \cdot T(e) = g_2$ and, similarly, $T(e) \cdot g_2 = g_2$ so that $T(e)$ is the identity of G_2.

(2) $\quad T(g) \cdot T(g^{-1}) = T(g \cdot g^{-1}) \quad T$ is an isomorphism
$\qquad\qquad\qquad\quad = T(e) \qquad\qquad$ substituting e for $g \cdot g^{-1}$

Also, $T(g^{-1}) \cdot T(g) = T(e)$ and since $T(e)$ is the identity of G_2, $T(g^{-1})$ is the inverse of $T(g)$.

Exercises

95. Obtain the right regular representation for the group $P_{12} = \{1, 5, 7, 11\}$.

212 Groups

1 5 7 11, 5 1 11 7

7 11 1 5, 11 7 5 1

I, (12)(34)

(13)(24), (14)(23)

$\{I, (12)(34), (13)(24), (14)(23)\}$

	1	5	7	11
	I	(12)(34)	(13)(24)	(14)(23)
1 I	1 I	5 (12)(34)	7 (13)(24)	11 (14)(23)
5 (12)(34)	5 (12)(34)	1 I	11 (14)(23)	7 (13)(24)
7 (13)(24)	7 (13)(24)	11 (14)(23)	1 I	5 (12)(34)
11 (14)(23)	11 (14)(23)	7 (12)(34)	5 (12)(34)	1 I

(a)

$$R_1 = \begin{pmatrix} 1 & 5 & 7 & 11 \\ \underline{\quad} & \underline{\quad} & \underline{\quad} & \underline{\quad} \end{pmatrix} \quad R_5 = \begin{pmatrix} 1 & 5 & 7 & 11 \\ \underline{\quad} & \underline{\quad} & \underline{\quad} & \underline{\quad} \end{pmatrix}$$

$$R_7 = \begin{pmatrix} 1 & 5 & 7 & 11 \\ \underline{\quad} & \underline{\quad} & \underline{\quad} & \underline{\quad} \end{pmatrix} \quad R_{11} = \begin{pmatrix} 1 & 5 & 7 & 11 \\ \underline{\quad} & \underline{\quad} & \underline{\quad} & \underline{\quad} \end{pmatrix}$$

(b) Write the elements R_1, R_2, R_3, R_4 in cycle notation. In this case it is more convenient to relabel the elements 1, 5, 7, 11 as 1, 2, 3, 4 respectively. Then, for example,

$$R_5 = \begin{pmatrix} 1 & 2 & 3 & 4 \\ 2 & 1 & 4 & 3 \end{pmatrix}$$

and in cycle notation $R_5 = (12)(34)$.

$R_1 = \underline{\qquad\qquad}$, $R_5 = \underline{\qquad\qquad}$

$R_7 = \underline{\qquad\qquad}$, $R_{11} = \underline{\qquad\qquad}$

Therefore, $G(R) = \underline{\qquad\qquad\qquad}$

(c) Complete the following group table.

	1	5	7	11
	I	(12)(34)	(13)(24)	(14)(23)
1 I				
5 (12)(34)				
7 (13)(24)				
11 (14)(23)				

96. Obtain the right regular representation for the group $P_{15} = \{1, 2, 4, 7, 8, 11, 13, 14\}$.

Groups   213

(a)

1 2 4 7 8 11 13 14

$$R_1 = \begin{pmatrix} 1 & 2 & 4 & 7 & 8 & 11 & 13 & 14 \\ & & & & & & & \end{pmatrix}$$

2 4 8 14 1 7 11 13

$$R_2 = \begin{pmatrix} 1 & 2 & 4 & 7 & 8 & 11 & 13 & 14 \\ & & & & & & & \end{pmatrix}$$

4 8 1 13 2 14 7 11

$$R_4 = \begin{pmatrix} 1 & 2 & 4 & 7 & 8 & 11 & 13 & 14 \\ & & & & & & & \end{pmatrix}$$

7 14 13 4 11 2 1 8

$$R_7 = \begin{pmatrix} 1 & 2 & 4 & 7 & 8 & 11 & 13 & 14 \\ & & & & & & & \end{pmatrix}$$

8 1 2 11 4 13 14 7

$$R_8 = \begin{pmatrix} 1 & 2 & 4 & 7 & 8 & 11 & 13 & 14 \\ & & & & & & & \end{pmatrix}$$

11 7 14 2 13 1 8 4

$$R_{11} = \begin{pmatrix} 1 & 2 & 4 & 7 & 8 & 11 & 13 & 14 \\ & & & & & & & \end{pmatrix}$$

13 11 7 1 14 8 4 2

$$R_{13} = \begin{pmatrix} 1 & 2 & 4 & 7 & 8 & 11 & 13 & 14 \\ & & & & & & & \end{pmatrix}$$

14 13 11 8 7 4 2 1

$$R_{14} = \begin{pmatrix} 1 & 2 & 4 & 7 & 8 & 11 & 13 & 14 \\ & & & & & & & \end{pmatrix}$$

(b) Write the elements $R_1, R_2, R_4, \ldots, R_{14}$ in cycle notation after the elements 1, 2, 4, 7, 8, 11, 13, 14 have been relabeled 1, 2, 3, 4, 5, 6, 7, 8 respectively.

I, (1235)(4876)

(13)(25)(47)(68), (1437)(2856)

(1734)(2658), (18)(27)(36)(45)

$R_1 = $ _____ , $R_2 = $ _____

$R_4 = $ _____ , $R_7 = $ _____

$R_{13} = $ _____ , $R_{14} = $ _____

(c) Complete the table on p. 214.

97. (a) Prove that every group is isomorphic to itself. (Isomorphism is reflexive.)
  (b) Prove that if a group G is isomorphic to a group K, then K is isomorphic to G. (Isomorphism is symmetric.)
  (c) Prove that if G, H and K are groups, G is isomorphic to H and H is isomorphic to K, then G is isomorphic to K. (Isomorphism is transitive.)

	1 *I*	2 (1235)(4876)	4 (13)(25)(47)(68)	7 (1437)(2856)	8 (1532)(4678)	11 (16)(24)(38)(57)	13 (1734)(2658)	14 (18)(27)(36)(45)
1 *I*								
2 (1235)(4876)								
4 (13)(25)(47)(68)								
7 (1437)(2856)								
8 (1532)(4678)								
11 (16)(24)(38)(57)								
13 (1734)(2658)								
14 (18)(27)(36)(45)								

	1 I	2 (1235)(4876)	4 (13)(25)(47)(68)	7 (1437)(2856)	8 (1532)(4678)	11 (16)(24)(38)(57)	13 (1734)(2658)	14 (18)(27)(36)(45)
1 I	1 I	2 (1235)(4876)	4 (13)(25)(47)(68)	7 (1437)(2856)	8 (1532)(4678)	11 (16)(24)(38)(57)	13 (1734)(2658)	14 (18)(27)(36)(45)
2 (1235)(4876)	2 (1235)(4876)	4 (13)(25)(47)(68)	8 (1532)(4678)	14 (18)(27)(36)(45)	1 I	13 (1734)(2658)	11 (16)(24)(38)(57)	7 (1437)(2856)
4 (13)(25)(47)(68)	4 (13)(25)(47)(68)	8 (1532)(4678)	1 I	13 (1734)(2658)	2 (1235)(4876)	7 (1437)(2856)	14 (18)(27)(36)(45)	11 (16)(24)(38)(57)
7 (1437)(2856)	7 (1437)(2856)	14 (18)(27)(36)(45)	13 (1734)(2658)	4 (13)(25)(47)(68)	11 (16)(24)(38)(57)	2 (1235)(4876)	1 I	8 (1532)(4678)
8 (1532)(4678)	8 (1532)(4678)	1 I	2 (1235)(4876)	11 (16)(24)(38)(57)	4 (13)(25)(47)(68)	14 (18)(27)(36)(45)	7 (1437)(2856)	13 (1734)(2658)
11 (16)(24)(38)(57)	11 (16)(24)(38)(57)	13 (1734)(2658)	7 (1437)(2856)	2 (1235)(4876)	14 (18)(27)(36)(45)	8 (1532)(4678)	4 (13)(25)(47)(68)	1 I
13 (1734)(2658)	13 (1734)(2658)	11 (16)(24)(38)(57)	14 (18)(27)(36)(45)	1 I	7 (1437)(2856)	4 (13)(25)(47)(68)	8 (1532)(4678)	2 (1235)(4876)
14 (18)(27)(36)(45)	14 (18)(27)(36)(45)	7 (1437)(2856)	11 (16)(24)(38)(57)	8 (1532)(4678)	13 (1734)(2658)	1 I	2 (1235)(4876)	4 (13)(25)(47)(68)

98. Prove that all groups of order 2 are isomorphic to each other.

99. Prove that all groups of order 3 are isomorphic to each other.

100. Prove that a group of order 4 is isomorphic to $C(4)$ or to the Klein Four-group (see Problem 47, p. 158).

101. Prove that a group G is commutative if and only if the following is true:
$$T: g \longrightarrow g^{-1} \quad \text{for all } g \in G$$
defines an isomorphic mapping of G onto G.

102. Suppose that G is a group and T is an isomorphism of G onto a group K. If $g \in G$, prove that g and $T(g)$ have the same order.

4.9 Homomorphism

We have seen above how two groups G and K can be related to each other by being isomorphic. In this case there is a 1–1 mapping T of G onto K such that
$$T(gh) = T(g)T(h) \tag{4-3}$$
for all $g, h \in G$. If the mapping T is only required to be a mapping of G onto K, not necessarily 1–1, and (4-3) is required to hold, then T is called a _homomorphism_ of G onto K.

To illustrate this concept, consider the groups $C(6)$ and $(Z_3, \oplus)$. We have $Z_3 = \{0, 1, 2\}$ and we may write $C(6) = \{e, g, g^2, g^3, g^4, g^5\}$. Now define the mapping T as follows:
$$T = \begin{pmatrix} e & g & g^2 & g^3 & g^4 & g^5 \\ 0 & 1 & 2 & 0 & 1 & 2 \end{pmatrix}$$

It is clear that T is a mapping of $C(6)$ onto Z_3. That the mapping T satisfies the equation (4-3) may be seen from the following table. Below each element of $C(6)$ we have placed the image in Z_3.

| | e | g | g^2 | g^3 | g^4 | g^5 |
	0	1	2	0	1	2
e	e	g	g^2	g^3	g^4	g^5
0	0	1	2	0	1	2
g	g	g^2	g^3	g^4	g^5	e
1	1	2	0	1	2	0
g^2	g^2	g^3	g^4	g^5	e	g
2	2	0	1	2	0	1
g^3	g^3	g^4	g^5	e	g	g^2
0	0	1	2	0	1	2
g^4	g^4	g^5	e	g	g^2	g^3
1	1	2	0	1	2	0
g^5	g^5	e	g	g^2	g^3	g^4
2	2	0	1	2	0	1

DEFINITION. Let T be a homomorphism of a group G onto a group H. Then
(1) H is called a *homomorphic image* of G;
(2) The subset of G $\{g \mid g \in G, T(g) = e\}$, where e is the identity of H, is called the *kernel* of T and is denoted by Ker(T). Ker(T) is illustrated in Fig. 4-6.

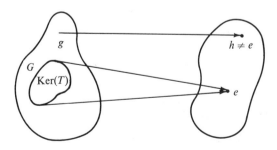

Figure 4-6

Lemma. If T is a homomorphism of a group G onto a group H, then Ker(T) is a normal subgroup of G.

Proof. (1) If e is the identity element of G, then by the lemma on p. 211, $T(e)$ is the identity element of H. By definition of Ker(T) we have $e \in$ Ker(T).

(2) If $a, b \in$ Ker(T), $T(a)$ and $T(b)$ are each the identity of H. Therefore, $T(a) \cdot T(b)$ is also the identity of H. But since T is a homomorphism, $T(a) \cdot T(b) = T(ab)$, which shows that $T(ab)$ is the identity of H. Hence, $ab \in$ Ker(T).

(3) If $a \in$ Ker(T), then $T(a)$ is the identity of H and by the lemma on p. 211, $T(a^{-1})$ is the inverse of $T(a)$. Since $T(a)$ is the identity of H and the identity is its own inverse, $T(a^{-1})$ is also the identity of H. It then follows that $a^{-1} \in$ Ker(T).

So far it has been shown that Ker(T) is a subgroup of G. To show that Ker(T) is a normal subgroup, we must show that $gag^{-1} \in$ Ker(T) for all $g \in G$ and $a \in$ Ker(T). Consider $T(gag^{-1})$. Since T is a homomorphism, we have

$$T(gag^{-1}) = T(g) \cdot T(a) \cdot T(g^{-1})$$

Now, if $a \in$ Ker(T), $T(a)$ is the identity of H. Then

$$\begin{aligned} T(gag^{-1}) &= T(g) \cdot T(a) \cdot T(g^{-1}) \\ &= T(g) \cdot T(g^{-1}) \end{aligned}$$

Therefore, $T(gag^{-1})$ is the identity element of H, since $T(g^{-1})$ is the inverse of $T(g)$. We then conclude that $gag^{-1} \in$ Ker(T) which proves that Ker(T) is normal in H.

Example 4-26: In the mapping on p. 216,

$$T = \begin{pmatrix} e & g & g^2 & g^3 & g^4 & g^5 \\ 0 & 1 & 2 & 0 & 1 & 2 \end{pmatrix}$$

is a homomorphism from $C(6)$ onto $(Z_3, \oplus)$. Since $T: e \to 0$ and $T: g^3 \to 0$ and 0 is the identity of Z_3,

$$\text{Ker}(T) = \{e, g^3\}$$

which is a normal subgroup of $C(6)$.

Example 4-27: Define a mapping T of the group of integers $(Z, +)$ onto the group $(\{1, -1\}, \cdot)$, where $\cdot$ is ordinary multiplication, as follows:

$$T: n \longrightarrow 1 \text{ if } n \text{ is even}$$
$$T: n \longrightarrow -1 \text{ if } n \text{ is odd}$$

For T to be a homomorphism we must have

$$T(n + m) = T(n) \cdot T(m) \tag{4-4}$$

for $n, m \in Z$. If both n, m are even, then $n + m$ is even and by definition of T

$$T(n + m) = 1, T(n) = 1, T(m) = 1$$

Thus (4-4) holds if both n, m are even. If both n, m are odd, $n + m$ is even so that

$$T(n + m) = 1, T(n) = -1, T(m) = -1$$

Again, we see that (4-4) holds. Suppose n is even and m is odd. Then $n + m$ is odd and
$$T(n + m) = -1, T(n) = 1, T(m) = -1$$
In this case, (4-4) also holds. The same is true if n is odd and m is even. Therefore, T is a homomorphism of $(Z, +)$ onto $(\{1, -1\}, \cdot)$. Here the kernel of T is the subgroup of even integers.

Theorem 42. If G is a group and H is a normal subgroup of G, then there is a homomorphism T of G onto the quotient group G/H and $\text{Ker}(T) = H$.

Proof. Define T as follows: for each $g \in G$
$$T: g \longrightarrow gH$$
This is clearly a mapping of G onto G/H. Also, we have for $a, b \in G$

$T(ab) = (ab)H$	by definition of T
$= (aH)(bH)$	by definition of coset multiplication
$= T(a) \cdot T(b)$	by definition of T

The identity element of G/H is H. This implies that the kernel of T is the set of all $g \in G$ such that $T(g) = H$. By definition of T, we have $T(g) = gH$. Therefore, if $T(g) = H$, $gH = H$ or $g \in H$. Thus, if $g \in \text{Ker}(T)$, $g \in H$; i.e., $\text{Ker}(T) \subseteq H$. Also, if $g \in H$, then $T(g) = gH = H$ which implies that $g \in \text{Ker}(T)$. We have shown that $\text{Ker}(T) \subseteq H$ and that $H \subseteq \text{Ker}(T)$; hence, $H = \text{Ker}(T)$. This proves the theorem.

Example 4-28: We have seen that the subgroup A_n is normal in S_n since $[S_n : A_n] = 2$. It then follows by Theorem 42 that S_n/A_n is a homomorphic image of S_n.

Theorem 42 produces a homomorphism T from a given normal subgroup and Theorem 41 is concerned with the converse situation: from a given homomorphism T a normal subgroup may be produced, namely $\text{Ker}(T)$.

Theorem 43. If T is a homomorphism of a group G onto a group K, then the quotient group $G/\text{Ker}(T)$ is isomorphic to K.

Proof. Denote $\text{Ker}(T)$ by H. Define Q as follows: for each $g \in G$
$$Q: Hg \longrightarrow T(g)$$
(1) *Q is a mapping*. Suppose that
$$Q: Ha \longrightarrow T(a)$$
$$Q: Hb \longrightarrow T(b)$$
and that $Ha = Hb$. Then $ab^{-1} \in H$ and since H is the kernel of T,
$$T(ab^{-1}) = e$$
where e is the identity of K. But since T is a homomorphism,
$$T(ab^{-1}) = T(a)T(b^{-1}) = e$$

Then, since $T(b^{-1})$ is the inverse of $T(b)$,
$$T(a)T(b^{-1})T(b) = e \cdot T(b)$$
$$T(a)e = T(b)$$
$$T(a) = T(b)$$

Therefore, Q is a mapping.

(2) *Q is a 1-1 mapping.* If $T(a) = T(b)$, the argument in part (1) may be reversed to show that $Ha = Hb$.

(3) *Q is a mapping onto K.* The mapping T is onto K. Therefore, if $k \in K$, there is an element $g \in G$ such that $T(g) = k$. Then by definition of Q
$$Q: \ Hg \longrightarrow T(g) = k$$
which shows that Q is onto.

(4) *Q is a homomorphism.*
$$\begin{aligned}Q(HaHb) &= Q(Hab) \\ &= T(ab) & \text{definition of } Q \\ &= T(a)T(b) & T \text{ is a homomorphism} \\ &= Q(Ha)Q(Hb) & \text{definition of } Q\end{aligned}$$

Therefore, Q is a homomorphism and this completes the proof that G/H is isomorphic to K.

Example 4-29: Consider the groups $(Z_8, \oplus)$ and $(Z_3^*, \otimes)$. We have
$$Z_8 = \{0, 1, 2, 3, 4, 5, 6, 7\}$$
and
$$Z_3^* = \{1, 2\}$$

The mapping T defined below is a homomorphism of Z_8 onto Z_3.
$$T = \begin{pmatrix} 0 & 1 & 2 & 3 & 4 & 5 & 6 & 7 \\ 1 & 2 & 1 & 2 & 1 & 2 & 1 & 2 \end{pmatrix}$$

The elements of Z_8 which map onto the identity of Z_3 are 0, 2, 4, 6. Hence, $\text{Ker}(T) = \{0, 2, 4, 6\}$. If we denote $\text{Ker}(T)$ by H, then the left cosets of H are H, $H1$. The following group tables are for the quotient group Z_8/H and Z_3.

$\otimes$	1	2
1	1	2
2	2	1

Z_3

	H	H
H	H	$H1$
$H1$	$H1$	H

Z_8/H

The relation between homomorphisms and normal subgroups, as indicated in Theorems 41 and 42, is a very important one. According to Theorem 42, if H is a normal subgroup of a group G, then H is the kernel of some homomorphism of G

onto a group. This result is sometimes helpful in testing a subgroup H to determine if it is normal.

Lemma. Let G be a group and let H be a subgroup of G such that $H \neq G$ and $H \neq \{e\}$. Suppose that for each homomorphic mapping T of G, we have:

$$H \subseteq \text{Ker}(T) \text{ implies } \text{Ker}(T) = G$$

Then H is not normal in G.

Proof. If H is a normal subgroup of G, then there is a homomorphism T of G such that $H = \text{Ker}(T)$. If $H \neq G$, then $\text{Ker}(T) \neq G$. Therefore, if H is normal, $H \neq \{e\}$ and $H \neq G$, it is not true that $H \subseteq \text{Ker}(T)$ implies $\text{Ker}(T) = G$.

Example 4-30: Consider the group S_3 and the subgroup $\{I, (12)\} = H$ of S_3. Suppose that T is a homomorphism of S_3 such that $H \subseteq \text{Ker}(T)$. We then have

$$T(I) = e \text{ and } T((12)) = e$$

where e is the identity element of the group onto which T maps S_3. The elements of S_3 are I, (12), (13), (23), (123), (132). We now wish to determine $T((13))$, $T((23))$, $T((123))$, $T((132))$. To do this we use the following equations:

(a) $(123) = (12)(123)(12)(132)$
(b) $(132) = (12)(132)(12)(123)$
(c) $(13) = (12)(123)$
(d) $(23) = (13)(123)$

Using (a) and the fact that T is a homomorphism, we have

(e) $T((123)) = T((12)(123)(12)(132))$ Using (a)
           $= T((12))T((123))T((12))T((132))$ T is a homomorphism
           $= e \cdot T((123)) \cdot e \cdot T((132))$ $T((12)) = e$
           $= T((123)) \cdot T((132))$
           $= T((123)(132))$ T is a homomorphism
           $= T(I)$
           $= e$ $T(I) = e$

Hence, $T((123)) = e$. Next,

$T((132)) = T((12)(132)(12)(123))$ Using (b)
          $= T((12))T((132))T((12))T((123))$ T is a homomorphism
          $= e \cdot T((132)) \cdot e \cdot T((123))$ $T((12)) = e$
          $= T((132))T((123))$
          $= T((132)(123))$ T is a homomorphism
          $= T(I)$
          $= e$ $T(I) = e$

(f) $T((13)) = T((12)(123))$ Using (e)
        $= T((12))T((123))$ T is a homomorphism
        $= e \cdot e$ Using $T((12)) = e$ and (e)
        $= e.$

$$T((23)) = T((13)(123))$$
$$= T((13))T((123))$$
$$= e \cdot e$$
$$= e$$

Using (d)
T is a homomorphism
Using (f)
and (e)

Thus, we have shown that Ker(T) = S_3 and, by the lemma above, $H = \{I, (12)\}$ is not a normal subgroup of S_3.

Exercises

$\{e\}, \{g, g^2, g^3, e\}$

$\{g^2, e\}, \{g^3, g^2, g, e\}$

$\{g\}$

$\{g^2\}$

$\{g^3\}$

103. Determine all the homomorphic images of the cyclic group $C(4) = \{e, g, g^2, g^3\}$.

    Solution. By Theorems 42 and 43, this may be solved by obtaining all normal subgroups of $C(4)$. But $C(4)$ is a commutative group, so that all subgroups are normal. Since $C(4)$ is cyclic, all the subgroups are cyclic. Hence, the subgroups are $[e], [g], [g^2], [g^3]$.

    $[e] = $ _____ , $[g] = $ _____

    $[g^2] = $ _____ , $[g^3] = $ _____

    The homomorphic images of $C(4)$ are now $C(4)/[e]$, $C(4)/[g]$, $C(4)[g^2]$, $C(4)/[g^3]$. To determine these quotient groups, the cosets must be computed.

    Left cosets of $[e] = H_1$:

    H_1

    $gH_1 = $ _____

    $g^2 H_1 = $ _____

    $g^3 H_1 = $ _____

    Left cosets of $[g] = H_2$:

    $H_2 (= C(4))$ is the only left coset of H_2.

    Left cosets of $[g^2] = H_3$:

$\{g^3, g\}$	H_3
	$gH_3 = $ _____
	Left cosets of $[g^3]$:
	These are the same as the left cosets of
$[g]$	_____
4	(a) The quotient group $C(4)/H_1$ has order ____
$C(4)$	Therefore, $C(4)/H_1$ is isomorphic to _____
the Klein 4-group	or _____. If an element of $C(4)/H_1$ can be found which will generate all of $C(4)/H_1$,
$C(4)$	then $C(4)H_1$ is isomorphic to _____
$\{gH_1, g^2H_1, g^3H_1, H_1\}$	Compute $[gH_1]$: $[gH_1] = $ _____
$C(4)/H_1$	Therefore, $[gH_1] = $ _____
	$C(4)/H_1$ is the cyclic group of order 4.
1	(b) The quotient group $C(4)/H_2$ has order ____
cyclic group of order 1	Therefore, $C(4)/H_2$ is the _____
2	(c) The quotient group $C(4)/H_3$ has order ____
$C(2)$	Hence, $C(4)/H_3$ is isomorphic to _____

(d) Since $[g^3] = [g]$, the quotient groups $C(4)/[g^3]$ and $C(4)/[g]$ are the same. From the preceding, we conclude that the homomorphic images of the cyclic group of order 4 are the cyclic groups of order 1, 2, 4.

104. Let $T: G \rightarrow K$ be a homomorphism of a group G onto a group K. Show by induction that if $g \in G$, then $T(g^n) = (T(g))^n$ for all $n \in Z$.

105. Show that every homomorphic image of a cyclic group is cyclic.

Solution. Let G be a cyclic group with generator g, $G = [g]$. Suppose that T is a homomorphism of G onto some group K. We wish to show that K is a _____ group. Let $k \in K$. Then since T is onto, there is an element $x \in G$ such that _____. But $G = [g]$, which implies that x may be written as _____, say $x = g^n$. Therefore, $T(x) = T(g^n)$, and from Problem 104, $T(g^n) = $ _____. Thus $k = (T(g))^n$, and we conclude that every element $k \in K$ can be expressed as a power of $T(g)$; i.e., K is generated by _____. This shows that K is a _____ group.

cyclic

$T(x) = k$

a power of g

$(T(g))^n$

$T(g)$

cyclic

106. Suppose G is a cyclic group generated by g and has finite order n. Show that if t is a positive integer such that $t \mid n$, then G has a subgroup of order t.

Solution. Since $G = [g]$ and has order n, we may write $G = $ _____.
If $t \mid n$, there is an integer m such that $n = $ _____.
Now consider the subgroup $[g^m]$. The order of $[g^m]$ is the same as the order of the element _____. Therefore, to determine the order of $[g^m]$, we determine the order of g^m. We have $(g^m)^t = g^{mt} = g^n = $ _____. It then follows that the order of g^m is not greater than _____. Next, suppose that $(g^m)^s = e$ for the integer s. Then $(g^m)^s = g^{ms} = e$. It then follows—

$\{e, g, g^2, \ldots, g^{n-1}\}$

tm

g^m

e

t

because n is the order of g—that $n \mid ms$. Thus there is some positive integer r such that _____ ($nr = ms$). Since $n = tm$, we may substitute in this last equation and obtain _____ ($tmr = ms$). Then cancelling the m gives _____ ($tr = s$). From the fact that r is a positive integer, it may be concluded that s is not less than _____ (t). This shows that no integer s which is less than t is such that $(g^m)^s = e$. Therefore, the order of g^m is _____ (t), and the order of $[g^m]$ is _____ (t).

107. Show that if $m \mid n$, the group $C(m)$ is a homomorphic image of $C(n)$.

Solution. We may write $n = mk$ for some positive integer k. By Problem 106, since $k \mid n$, $C(n)$ has a subgroup of order _____ (k), namely $C(k)$. The group $C(n)$ is commutative so that every subgroup is _____ (normal) in $C(n)$. In particular, $C(k)$ is normal in $C(n)$. By Theorem 42, the quotient group $C(n)/C(k)$ is a _____ _____ (homomorphic image) of $C(n)$. The order of the quotient group $C(n)/C(k)$ is $[C(n):C(k)]$. Since $n = mk$, $[C(n):C(k)] =$ _____ (m). Therefore, $C(n)$ has a homomorphic image of order m. By Problem 105, every homomorphic image of a cyclic group is cyclic. Therefore, the quotient group $C(n)/C(k)$ is a cyclic group of order m. This proves that $C(m)$ is a homomorphic image of $C(n)$.

108. Use the result of Problem 107 to determine the homomorphic images of the following groups.

(a) The homomorphic images of $C(6)$ are:

$C(1), C(2), C(3), C(6)$

(b) The homomorphic images of $C(8)$ are:

$C(1), C(2), C(4), C(8)$

(c) The homomorphic images of $C(14)$ are:

$C(1), C(2), C(7), C(14)$

(d) The homomorphic images of $C(36)$ are:

$C(1), C(2), C(4), C(3), C(9), C(6),$
$C(12), C(18), C(36)$

(e) $C(p)$, where p is prime:

$C(1), C(p)$

(f) $C(p^2)$, where p is prime:

$C(1), C(p), C(p^2)$

(g) $C(p \cdot q)$, where p and q are prime:

$C(1), C(p), C(q), C(pq)$

109. Let $H = \{I, (13)\}$ in the group S_3. (a) Show that H is not normal in S_3 by finding an element $g \in S_3$ such that $gH \neq Hg$. (b) Show that H is not normal in S_3 by showing that if $H \subseteq \text{Ker}(T)$, for some homomorphism T, then $\text{Ker}(T) = S_3$.

4.10 Direct Products

Suppose we are given two groups (G_1, B_1) and (G_2, B_2). From these two groups we may construct another group, called their *direct product*, as follows. Let $G = G_1 \times G_2$. On the set G—which is a set of ordered pairs—a binary operation B is defined: for $(a, b) \in G, (c, d) \in G$

$$(a, b)B(c, d) = (aB_1c, bB_2d) \tag{4-5}$$

As an example of this, consider the groups $(Z_3^*, \otimes)$ and $(Z_3, \oplus)$. In this case, $G_1 = Z_3^*$, $G_2 = Z_3$, $B_1 = \otimes$ and $B_2 = \oplus$. Since $Z_3^* = \{1, 2\}$ and $Z_3 = \{0, 1, 2\}$, $G = Z_3^* \times Z_3^* = \{(1, 0), (1, 1), (1, 2), (2, 0), (2, 1), (2, 2)\}$. The definition of B above is applied as follows, for example,

$$(1, 2)B(2, 1) = (1 \otimes 2, 2 \oplus 1)$$
$$= (2, 0)$$

From the tables for $(Z_3^*, \otimes)$ and $(Z_3, \oplus)$, the table for (G, B) is easily constructed.

$\otimes$	1	2
1	1	2
2	2	1

$\oplus$	0	1	2
0	0	1	2
1	1	2	0
2	2	0	1

B	(1, 0)	(1, 1)	(1, 2)	(2, 0)	(2, 1)	(2, 2)
(1, 0)	(1, 0)	(1, 1)	(1, 2)	(2, 0)	(2, 1)	(2, 2)
(1, 1)	(1, 1)	(1, 2)	(1, 0)	(2, 1)	(2, 2)	(2, 0)
(1, 2)	(1, 2)	(1, 0)	(1, 1)	(2, 2)	(2, 0)	(2, 1)
(2, 0)	(2, 0)	(2, 1)	(2, 2)	(1, 0)	(1, 1)	(1, 2)
(2, 1)	(2, 1)	(2, 2)	(2, 0)	(1, 1)	(1, 2)	(1, 0)
(2, 2)	(2, 2)	(2, 0)	(2, 1)	(1, 2)	(1, 0)	(1, 1)

Lemma. If (G_1, B_1) and (G_2, B_2) are groups, then (G, B) is a group, where $G = G_1 \times G_2$ and B is defined by (4-5).

Proof. (a) B *is a mapping of* $G \times G$ *into* G. If $(a, b) \in G$ and $(c, d) \in G$, then $a, c \in G_1$ and $b, d \in G_2$. Hence, $aB_1c \in G_1$ and $bB_2d \in G_2$ so that $(aB_1c, bB_2d) \in G$.

(b) (G, B) *has an identity element.* If e_1 is the identity of G_1 and e_2 is the identity of G_2, then (e_1, e_2) is the identity of G. For if $(a, b) \in G$, then

$$(a, b)B(e_1, e_2) = (aB_1e_1, bB_2e_2) = (a, b)$$

and

$$(e_1, e_2)B(a, b) = (e_1B_1a, e_2B_2b) = (a, b)$$

(c) B *is associative.* If $(a, b), (c, d), (f, g) \in G$, then

$$[(a, b)B(c, d)]B(f, g) = (aB_1c, bB_2d)B(f, g)$$
$$= ((aB_1c)B_1f, (bB_2d)B_2g)$$

and
$$(a, b)B[(c, d)B(f, g)] = (a, b)B(cB_1f, dB_2g)$$
$$= (aB_1(cB_1f), bB_2(dB_2g))$$

Now recall that for ordered pairs (x, y) and (z, w) we have $(x, y) = (z, w)$ if and only if $x = z$ and $y = w$. Then since B_1 and B_2 are associative
$$(aB_1c)B_1f = aB_1(cB_1f)$$
and
$$(bB_2d)B_2g = bB_2(dB_2g)$$

Hence, $((aB_1c)B_1f, (bB_2d)B_2g) = (aB_1(cB_1f), bB_2(dB_2g))$ and it follows that B is associative.

(d) *Each element of G has an inverse in G*. If $(a, b) \in G$, then $(a^{-1}, b^{-1}) \in G$ is the inverse of (a, b); i.e. $(a, b)^{-1} = (a^{-1}, b^{-1})$:
$$(a, b)B(a^{-1}, b^{-1}) = (aB_1a^{-1}, bB_2b^{-1}) = (e_1, e_2)$$
and
$$(a^{-1}, b^{-1})B(a, b) = (a^{-1}B_1a, b^{-1}B_2b) = (e_1, e_2)$$

This proves the lemma.

Usually the binary operations are not distinguished in two groups when the direct product is considered. For example, the groups may be given as $(G_1, \cdot)$, $(G_2, \cdot)$ or simply as G_1, G_2. To add to the confusion, B is also denoted by $\cdot$ so that (4-5) becomes
$$(a, b) \cdot (c, d) = (a \cdot c, b \cdot d) \tag{4-6}$$

In (4-6) the symbol "$\cdot$" is used to denote the three different binary operations. If G_1, G_2 are groups, then the Cartesian product notation "$G_1 \times G_2$" is also used to denote the direct product group.

Theorem 44. If G_1, G_2 are groups, the direct product $G_1 \times G_2$ contains a normal subgroup which is isomorphic to G_1 and also contains a normal subgroup which is isomorphic to G_2.

Proof. If e_2 is the identity element of G_2, define G_1^* as follows:
$$G_1^* = \{(g, e_2) | g \in G_1\}$$
G_1^* is a normal subgroup of $G_1 \times G_2$.

(1) If $(g, e_2) \in G_1$ and $(h, e_2) \in G_1$, then
$$(g, e_2) \cdot (h, e_2) = (hg, e_2 \cdot e_2) = (hg, e_2) \in G_1^*$$

(2) The identity element (e_1, e_2) of $G_1 \times G_2$ is in G_1^* by definition of G_1^*.

(3) The inverse of each element (g, e_2) of G_1^* is (g^{-1}, e_2) which is in G_1^*.
(4) To show that G_1^* is normal in $G_1 \times G_2$, we must show that

$$(a, b) \cdot (g, e_2) \cdot (a, b)^{-1} \in G_1^*$$

for all $(a, b) \in G_1 \times G_2$ and $(g, e_2) \in G_1^*$. But this is immediate since

$$\begin{aligned}(a, b) \cdot (g, e_2) \cdot (a, b)^{-1} &= (a, b) \cdot (g, e_2) \cdot (a^{-1}, b^{-1}) \\ &= (aga^{-1}, be_2 b^{-1}) \\ &= (aga^{-1}, e_2) \in G_1^*\end{aligned}$$

This shows that G_1^* is a normal subgroup of $G_1 \times G_2$. To show that G_1^* is isomorphic to G_1, define the mapping T as follows:

$$T: g \longrightarrow (g, e_2)$$

for all $g \in G_1$. This is clearly a mapping of G_1 onto G_1^*. It is also 1–1, since if $(g, e_2) = (h, e_2)$, then $g = h$. Finally, if $g, h \in G_1$, then

$$\begin{aligned}T(g \cdot h) &= (g \cdot h, e_2) & \text{by definition of } T \\ &= (g, e_2) \cdot (h, e_2) \\ &= T(g) \cdot T(h) & \text{by definition of } T\end{aligned}$$

Therefore, T is an isomorphism. In a similar fashion, the set $G_2^* = \{(e_1, c) \mid c \in G_2\}$ is a normal subgroup of $G_1 \times G_2$, which is isomorphic to G_2. This completes the proof.

The direct product concept may be extended to a consideration of more than two groups. For example, if G_1, G_2, G_3 are groups, we may form the set $G_1 \times G_2 \times G_3$ of all ordered triples (g_1, g_2, g_3) and define a binary operation on this set in a manner similar to (4-5). We again obtain a group; this group contains three normal subgroups isomorphic to G_1, G_2, G_3 respectively.

We now wish to catalogue all the groups of orders 1, 2, 3, 4, 5, 6, 7. By Theorem 34, p. 186 we know that groups of prime order are cyclic and by Theorem 38, p. 207, cyclic groups of the same order are isomorphic. Therefore, the only groups of orders 1, 2, 3, 5, 7 are $C(1), C(2), C(3), C(5), C(7)$ respectively. It remains then to determine the groups of orders 4 and 6. In a group of order 4, an element $g \neq e$ must have order 2 or 4. If there is an element of order 4, then the group is the cyclic group $C(4)$. If there is no element of order 4, then every element other than the identity has order 2. Let G be a group of order 4 with no element of order 4. Then if $a \in G$, $a \neq e$ we have $a^2 = e$. There are in G two elements other than e, a. Let $b \in G$, $b \neq e$ and $b \neq a$. Therefore, we have the three elements e, a, b where $a^2 = e$ and $b^2 = e$. The element $a \cdot b \in G$, and we cannot have $ab = e$, $ab = a$ or $ab = b$. For if $ab = e$, then $(ab)b = eb$ or $a = b$; if $ab = a$, then $a(ab) = a^2$ or $b = e$; if $ab = b$, then $(ab)b = b^2$ or $a = e$. G consists then of the four elements e, a, b, ab. Since $ba \in G$, ba must be one of the elements e, a, b, ab. By these arguments we may show that $ba \neq e$, $ba \neq a$, $ba \neq b$. Hence, $ab = ba$. Now the group table for G may be constructed.

	e	a	b	ab
e	e	a	b	ab
a	a	e	ab	b
b	b	ab	e	a
ab	ab	b	a	e

In this group table, the entry for $b(ab)$ was computed by using $ab = ba$: $b(ab) = b(ba) = b^2a = ea = a$. Also, $(ab)a = ba^2 = be = b$; $(ab)(ab) = (ba)(ab) = ba^2b = beb = b^2 = e$. This group of order 4 is isomorphic to the Klein Four-group P_{12}. P_{12} is $\{1, 5, 7, 11\}$ with the following table:

$\otimes$	1	5	7	11
1	1	5	7	11
5	5	1	11	7
7	7	11	1	5
11	11	7	5	1

It is easily verified that the following mapping, T, is an isomorphism from P_{12} onto $\{e, a, b, ab\}$.

$$T = \begin{pmatrix} 1 & 5 & 7 & 11 \\ e & a & b & ab \end{pmatrix}$$

For example,

$$T(5 \times 7) = T(11) = ab$$
$$T(5) = a$$
$$T(7) = b$$

so that

$$T(5 \times 7) = T(5)T(7)$$

From the above considerations it follows that there are exactly two groups of order 4, namely $C(4)$ and the Klein Four-group.

Before determining the groups of order 6, we need the following lemmas.

Lemma 1. Let G be a group and let $a \in G$, $b \in G$. If $ab = ba$, then for all $m, n \in Z$,

(1) $a^n b = ba^n$

(2) $a^n b^m = b^m a^n$
(3) $(ab)^n = a^n b^n$

Proof. (1) First, use induction to prove the lemma for all positive integers n. If $n = 1$, then $a^n b = ba^n$ is true, since we are given that $ab = ba$. Now suppose this is true for k; i.e., suppose that $a^k b = ba^k$. Then

$$\begin{aligned} a^{k+1} b &= a^k \cdot ab && \text{by a law of exponents} \\ &= a^k \cdot ba && \text{since } ab = ba \\ &= (a^k b) a \\ &= (ba^k) a && \text{since } a^k b = ba^k \\ &= ba^{k+1} && \text{by a law of exponents} \end{aligned}$$

Thus (1) is true for all positive integers. For $n = 0$, part (1) is obviously true. If n is a positive integer, we have

$$a^{-n} b = (a^{-1})^n b$$

by definition of a^{-n}. Now, the hypothesis is that $ab = ba$ and from this

$$\begin{aligned} a^{-1} ab &= a^{-1} ba \\ eb &= a^{-1} ba \\ b &= a^{-1} ba \end{aligned}$$

But from $b = a^{-1} ba$, it follows that $ba^{-1} = a^{-1} baa^{-1} = a^{-1} b$. We may then apply (1) to the elements a^{-1}, b and the positive integer n to obtain $(a^{-1})^n b = b(a^{-1})^n$. Therefore,

$$\begin{aligned} a^{-n} b &= (a^{-1})^n b \\ &= b(a^{-1})^n \\ &= ba^{-n} \end{aligned}$$

i.e., $a^{-n} b = ba^{-n}$ so that (1) holds for all integers.

(2) To prove that $a^n b^m = b^m a^n$ we first apply (1) to the elements a, b and the integer m to obtain $ab^m = b^m a$. Then, since $ab^m = b^m a$, the hypothesis of the lemma is again satisfied by the two elements a, b^m so that (1) may be applied to a, b^m and the integer n to obtain $a^n b^m = b^m a^n$.

(3) For positive integers, we again use induction. If $n = 1$, $(ab)^1 = a^1 b^1$ is true by definition. Consider $(ab)^{k+1}$:

$$\begin{aligned} (ab)^{k+1} &= (ab)^k \cdot (ab) \\ &= (a^k b^k) \cdot (ab) && \text{if } (ab)^k = a^k b^k \\ &= a^k \cdot (b^k \cdot a) \cdot b \\ &= a^k \cdot (ab^k) \cdot b && \text{by (1)} \\ &= (a^k \cdot a) \cdot (b^k \cdot b) \\ &= a^{k+1} \cdot b^{k+1} \end{aligned}$$

Therefore, if (3) is true for k, it is true for $k + 1$. This is also true for 0 since $(ab)^0 = e$

and $a^0 b^0 = e$. For negative integers we have: if $n > 0$,

$$
\begin{aligned}
(ab)^{-n} &= ((ab)^{-1})^n && \text{by definition of } (ab)^{-n} \\
&= ((ba)^{-1})^n && \text{since } ab = ba \\
&= (a^{-1} b^{-1})^n && \text{since } (ba)^{-1} = a^{-1} b^{-1} \\
&= (a^{-1})^n \cdot (b^{-1})^n && \text{the result holds for positive integers} \\
&&& \text{and the elements } a^{-1}, b^{-1} \\
&= a^{-n} \cdot b^{-n} && \text{by definition}
\end{aligned}
$$

Hence, (3) holds for all integers, completing the proof of the lemma.

Lemma 2. Let G be a group, $a \in G$, $b \in G$ and let a, b have order n, m respectively. If n, m are relatively prime and if $ab = ba$, then ab has order $n \cdot m$.

Proof. The order of ab is not greater than $n \cdot m$, as the following computations show:

$$
\begin{aligned}
(ab)^{n \cdot m} &= a^{n \cdot m} b^{n \cdot m} && \text{by Lemma 1 (3), since } ab = ba \\
&= (a^n)^m (b^m)^n \\
&= e^m \cdot e^n && \text{since } a^n = e, b^m = e \\
&= e \cdot e \\
&= e
\end{aligned}
$$

Let t be the order of ab. Then $(ab)^t = e$, and therefore $(ab)^{nt} = e$, $(ab)^{mt} = e$.

$$
\begin{aligned}
(ab)^{nt} &= a^{nt} \cdot b^{nt} && \text{by Lemma 1 (3), since } ab = ba \\
&= (a^n)^t \cdot b^{nt} \\
&= e^t \cdot b^{nt} && \text{since } a^n = e \\
&= e \cdot b^{nt} \\
&= b^{nt} \\
(ab)^{mt} &= a^{mt} \cdot b^{mt} && \text{by Lemma 1 (3), since } ab = ba \\
&= a^{mt} \cdot (b^m)^t \\
&= a^{mt} \cdot e && \text{since } b^m = e \\
&= a^{mt} \cdot e \\
&= a^{mt}
\end{aligned}
$$

Hence, $b^{nt} = e$ and $a^{mt} = e$. Since n is the order of a and m is the order of b, it follows that $n \mid mt$ and $m \mid nt$. But n, m are relatively prime so that $n \mid t$, $m \mid t$ and $n \cdot m \mid t$. We conclude that $n \cdot m$ is the smallest positive integer such that $(ab)^{n \cdot m} = e$ and hence ab has order $n \cdot m$.

Lemma 3. If G is a group such that $g^2 = e$ for all $g \in G$, then G is Abelian.*

Proof. We must show that $xy = yx$ for all $x, y \in G$. By hypothesis, $(xy)^2 = e$; i.e.,

$$(xy)(xy) = e$$

---

* A group G is *Abelian* or *commutative* if $xy = yx$ for all $x, y \in G$.

From this it follows that
$$x(xy)(xy) = xe$$
$$x^2(y\,x\,y) = x$$
$$e(y\,x\,y) = x \quad \text{since } x^2 = e$$
$$y\,x\,y = x$$
$$y(y\,x\,y) = yx$$
$$y^2(xy) = yx$$
$$e(xy) = yx \quad \text{since } y^2 = e$$
$$xy = yx$$

Therefore, $xy = yx$ for all $x, y \in G$, and the lemma is proved.

Lemma 4. *If G is a group of order 6, then G contains an element of order 3.*

Proof. Since G has order 6, if $g \in G$ and $g \neq e$, then g has order 2 or order 3. If G has no element of order 3, then $g^2 = e$ for all $g \in G$. Then by Lemma 3, G is Abelian so that every subgroup of G is normal in G. Now let $g \in G$, $g \neq e$ and let $[g] = H$. Since g has order 2, $H = \{e, g\}$ and since G has order 6, $[G:H] = 3$. Therefore, H is a normal subgroup of G and G/H has order 3. But every group of order 3 is isomorphic to $C(3)$:

$$G/H \cong C(3)$$

This implies that G/H contains 3 cosets and there is an element $a \in G$, $a \notin H$ such that aH generates G/H. Thus aH has order 3:

$$(aH)^3 = H$$

or

$$a^3 H = H$$

But since we assume that $g^2 = e$ for all $g \in G$, $a^3 = a \cdot a^2 = a \cdot e = a$. Therefore, from $a^3 H = H$ we have

$$aH = H$$

From this it follows that $a \in H$, which is a contradiction. Hence, it is not true that $g^2 = e$ for all $g \in G$. There must, then, exist at least one element of order 3 in G.

Lemma 5. *If G is a group of order 6, then G contains an element of order 2.*

Proof. If $g \in G$, $g \neq e$, then g has order 2 or order 3. Suppose that G has no element of order 2. Then $g^3 = e$ for all $g \in G$. Let g be an element of order 3 and let $H = [g]$. Since H has order 3 and G has order 6, $[G:H] = 2$. Therefore, H is normal in G by the lemma on p. 195. Then the quotient group G/H has order 2 and is isomorphic to $C(2)$. Thus G/H has 2 cosets, say H and Ha for some $a \in G$, $a \notin H$. Then $(Ha)^2 = H$ or $Ha^2 = H$. Now $H = \{e, g, g^2\}$ and $Ha^2 = \{a^2, ga^2, g^2a^2\}$. Hence, g^2a^2 must be one of the elements e, g, g^2. If $g^2a^2 = g^2$, then $a^2 = e$, contrary to the

assumption that G has no element of order 2. If $g^2a^2 = g$, then $ga^2 = e$ and, since a has order 3, $g = a$. Hence, $H \cap Ha \neq \emptyset$, which is a contradiction. Finally, if $g^2a^2 = e$, $g^2 = a$ and again $H \cap Ha \neq \emptyset$. We have obtained a contradiction in each of the cases $g^2a^2 = g^2$, $g^2a^2 = g$, $g^2a^2 = e$ and, therefore, conclude that G cannot have all elements $\neq e$ of order 3.

It has now been shown by Lemma 4 and Lemma 5 that a group G of order 6 must have at least one element of order 2 and one element of order 3. Let a and b be elements of G and of order 2 and 3 respectively. Then the elements e, a, b, b^2 are four of the elements of G. The product $ab \in G$ differs from each of e, a, b, b^2. For if $ab = e$, then $b = a$; if $ab = a$, then $b = e$; if $ab = b$, then $a = e$; if $ab = b^2$, then $a = b$. Thus five of the elements of G are e, a, b, b^2, ab. We may argue that $ba \in G$ is also different from e, a, b, b^2. Can we have $ab = ba$? By Lemma 2, the order of ab is 6 if $ab = ba$, since a has order 2 and b has order 3. In this case G is the group $C(6)$. Now suppose G has no element of order 6. Then $ab \neq ba$. The six elements of G must then be e, a, b, b^2, ab, ba. Now let us construct a table for G. Many entries in the table may be obtained by using group properties and the orders of a, b. For example, $a(ab) = a^2b = e \cdot b = b$. Thus in our table the entry for $a(ab)$ is b. Consider the problem of completing the table below from what we have.

	e	a	b	b^2	ab	ba
e	e	a	b	b^2	ab	ba
a	a	e	ab		b	
b	b	ba	b^2	e		
b^2	b^2		e	b		a
ab	ab			a		
ba	ba	b			b^2	

In the a-row, the two vacant entries must be either b^2 or ba, and since we cannot have repetition in rows or columns, it must be that

$$ab^2 = ba \tag{4-7}$$

and

$$aba = b^2 \tag{4-8}$$

We may derive the following equations from (4-7) and (4-8), and thereby complete the table.

$$b(ab) = (ba)b$$
$$= (ab^2)b \quad [\text{by (4-7)}]$$

$$= ab^3$$
$$= a$$

$$bab = a \tag{4-9}$$

$$b^2a = (aba)a \quad \text{[by (4-8)]}$$
$$= aba^2$$
$$= ab$$

$$b^2a = ab \tag{4-10}$$

$$b^2(ab) = b(bab)$$
$$= ba \quad \text{[by (4-9)]}$$

$$b^2(ab) = ba \tag{4-11}$$

$$(ab)a = (b^2a)a \quad \text{[by (4-10)]}$$
$$= b^2a^2$$
$$= b^2$$

$$aba = b^2 \tag{4-12}$$

$$ab^2 = a(aba) \quad \text{[by (4-12)]}$$
$$= a^2ba$$
$$= ba$$

$$ab^2 = ba \tag{4-13}$$

$$(ab)^2 = (ab)(ab)$$
$$= a(ba)b$$
$$= a(ab^2)b \quad \text{[by (4-13)]}$$
$$= a^2b^3$$
$$= e$$

$$(ab)^2 = e \tag{4-14}$$

$$(ab)(ba) = ab^2a$$
$$= a(aba)a \quad \text{[by (4-12)]}$$
$$= a^2ba^2$$
$$= b$$

$$ab^2a = b \tag{4-15}$$

$$bab^2 = (bab)b$$
$$= ab \quad \text{[by (4-9)]}$$

$$bab^2 = ab \tag{4-16}$$

$$(ba)^2 = (ba)(ba)$$
$$= (ab^2)(ba) \quad [\text{by (4-13)}]$$
$$= ab^3 a$$
$$= a^2$$
$$= e$$

$$(ba)^2 = e \tag{4-17}$$

We may now complete the table for G.

	e	a	b	b^2	ab	ba
e	e	a	b	b^2	ab	ba
a	a	e	ab	ba	b	b^2
b	b	ba	b^2	e	a	ab
b^2	b^2	ab	e	b	ba	a
ab	ab	b^2	ba	a	e	b
ba	ba	b	a	ab	b^2	e

This is the table of a group which is isomorphic to $S_3 = \{I, (12), (13), (23), (123), (132)\}$. The following mapping T is an isomorphism:

$$T = \begin{pmatrix} I & (12) & (13) & (123) & (23) & (132) \\ e & a & ab & b & ba & b^2 \end{pmatrix}$$

From these considerations it follows that there are two groups of order 6, namely, $C(6)$ and S_3.

Exercises

110. Determine all the subgroups of $(Z, +)$.

111. Determine all the homomorphic images of the group $(Z, +)$.

112. Construct group tables for $C(2)$ and $C(3)$, then construct the table for the direct product $C(2) \times C(3)$. Show that $C(2) \times C(3)$ is isomorphic to $C(6)$.

113. Construct group tables for $C(2)$ and $C(5)$, then construct the table for the direct product $C(2) \times C(5)$. Show that $C(2) \times C(5)$ is isomorphic to $C(10)$.

114. If n, m are relatively prime natural numbers, show that $C(n) \times C(m)$ is isomorphic to $C(n \cdot m)$.

115. (a) Show that the group $P_{15} = \{1, 2, 4, 7, 8, 11, 13, 14\}$ is not cyclic. (Recall that the operation is multiplication modulo 15.)
 (b) Compute the subgroups [2] and [11] of P_{15}.
 (c) Show that P_{15} is isomorphic to $C(4) \times C(2)$.

116. (a) Show that the group P_{21} is not cyclic.
 (b) Show that P_{21} is isomorphic to $C(2) \times C(6)$.

117. In the symmetric group S_8, let
$$p = (1234)(5678) \quad q = (1537)(2846)$$
Use the table method to compute $[p, q]$. This is a group Q of order 8, known as the *quaternion* group.
 (a) Compute all the subgroups of the quaternion group.
 (b) Show that all the subgroups of the quaternion group are normal in Q.

The Theorem of Lagrange asserts that the order of a subgroup divides the order of the group. The converse of this result is not true, i.e., there exist groups G of order n and integers t such that $1 < t < n$ and $t \mid n$, but G contains no subgroup of order t. The following Exercises, 118–120, are designed to demonstrate this fact. These exercises show that the group A_4 of order 12 has no subgroup of order 6.

| 3 |
| 2 |
| |
| e |
| e |
| 2, 3 |

118. The group A_4 has been exhibited on p. 204. Observe that every element of A_4 is either a _____ cycle or a product of two disjoint _____ cycles. From this we conclude that if $g \in A_4$, then $g = (abc)$ or $g = (ab)(cd)$, where a, b, $d \in \{1, 2, 3, 4\}$. Then if $g = (abc)$, $g^3 = $ _____ and if $g = (ab)(cd)$, $g^2 = $ _____. Hence every element $g \in A_4$ such that $g \neq e$ has order _____ or order _____.

119. Suppose that H is a normal subgroup of A_4 and

238   Groups

normal	that H contains a 3-cycle, i.e., some element (abc). Then by definition of _____ subgroup, H also contains all elements of the form $g(abc)g^{-1}$ for all $g \in$ _____.
A_4	In particular, H contains each of the following elements:
	$(ab)(abc)(ab)^{-1} =$ _____
(acb)	$(ad)(abc)(ad)^{-1} =$ _____
(bcd)	$(bd)(abc)(bd)^{-1} =$ _____
(adc)	Also, H contains the inverse of each of these elements:
	$(abc)^{-1} =$ _____
(acb)	$(acb)^{-1} =$ _____
(abc)	$(bcd)^{-1} =$ _____
(bdc)	$(adc)^{-1} =$ _____
(acd)	Therefore, H contains the following elements:
$I, (abc), (acb), (bcd), (bdc), (acd), (adc)$	_____
12	Now A_4 has order _____ and H is a
7	subgroup of A_4 containing at least _____
12	elements. Hence, H must have order _____
A_4	and we conclude that $H =$ _____. This shows that the only normal subgroup of A_4 which contains a 3-cycle is _____
A_4	
	120. Suppose that A_4 has a subgroup H of order 6. Then, since A_4 has order _____
12	$[A_4:H] =$ _____. But if H has index 2 in A_4, it follows that H is a _____
2	
normal	

subgroup of A_4. By Lemma 4, a group of order 6 contains an element of order ___3___. Therefore, H contains an element of order 3 or a ___3___-cycle. Thus H is a normal subgroup of A_4 which contains a 3-cycle. This implies—by Exercise 119—that $H = $ ___A_4___.

But this is a ___contradiction___ since H was assumed to have order ___6___. We may then conclude that A_4 contains no subgroup of order 6.

Exercises 121–136 contain a discussion of the concepts *conjugate, normalizer, center*. The definitions of these terms follow.

Definition. Let G be a group.

(a) If $a \in G$ and $b \in G$, then a is *conjugate* to b if there is some $x \in G$ such that $a = xbx^{-1}$. If a is conjugate to b, we write aCb.

(b) If $g \in G$, then $\{y \mid y \in G \text{ and } gy = yg\}$ is called the *normalizer* of g in G and is denoted by $N(g)$.

(c) The *center* of G is $\{z \mid zg = gz \text{ for all } g \in G\}$.

121. Let G be the group $C(4) = \{e, a, a^2, a^3\}$. The conjugates of a^2 may be computed as follows:

$ea^2e^{-1} = $ ___ea^2e___ $= $ ___a^2___

$aa^2a^{-1} = $ ___aa^2a^3___ $= $ ___a^2___

$a^2a^2a^{-2} = $ ___a^2e___ $= $ ___a^2___

$a^3a^2a^{-3} = $ ___a^3a^2a___ $= $ ___a^2___

From these calculations we may conclude that

	that the only element of $C(4)$ which is conjugate to a^2 is _____
a^2	Now compute the conjugates of a:
	$eae^{-1} =$ _____
a	$aaa^{-1} =$ _____
a	$a^2aa^{-2} =$ _____
a	$a^3aa^{-3} =$ _____
a	From these computations we conclude that the only conjugate of a is _____
a	The group $C(4)$ is Abelian, so that for all g, $x \in C(4)$ we have $xgx^{-1} =$ _____
$xx^{-1}g$	
eg, g	= _____ = _____
	From this it follows that for any $g \in C(4)$ the only conjugate of g is _____
g	

122. Let G be the group $S_3 = \{I, (12), (13), (23), (123), (132)\}$.

(a) Compute the conjugates of (12).

$I(12)I, (12)$	$I(12)I^{-1} =$ _____ = _____
$(12)(12)(12), (12)$	$(12)(12)(12)^{-1} =$ _____ = _____
$(13)(12)(13), (23)$	$(13)(12)(13)^{-1} =$ _____ = _____
$(23)(12)(23), (13)$	$(23)(12)(23)^{-1} =$ _____ = _____
$(123)(12)(132), (13)$	$(123)(12)(123)^{-1} =$ _____ = _____
$(132)(12)(123), (23)$	$(132)(12)(132)^{-1} =$ _____ = _____

Therefore, the conjugates of (12) in S_3 are

(12), (23), (13)   _____

(b) Compute the conjugates of (13):

(13) $I(13)I^{-1} =$ _____

Groups   241

(12)(13)(12), (23)	$(12)(13)(12)^{-1} =$ _____	= _____
(13)(13)(13), (13)	$(13)(13)(13)^{-1} =$ _____	= _____
(23)(13)(23), (12)	$(23)(13)(23)^{-1} =$ _____	= _____
(123)(13)(132), (23)	$(123)(13)(123)^{-1} =$ _____	= _____
(132)(13)(123), (12)	$(132)(13)(132)^{-1} =$ _____	= _____

Therefore, the conjugates of (13) in S_3 are

(12), (13), (23) _____

(c) Compute the conjugates of (23) in S_3:

(23) $I(23)I^{-1} =$ _____

(12)(23)(12), (13)	$(12)(23)(12)^{-1} =$ _____	= _____
(13)(23)(13), (12)	$(13)(23)(13)^{-1} =$ _____	= _____
(23)(23)(23), (23)	$(23)(23)(23)^{-1} =$ _____	= _____
(123)(23)(132), (12)	$(123)(23)(123)^{-1} =$ _____	= _____
(132)(23)(123), (13)	$(132)(23)(132)^{-1} =$ _____	= _____

Therefore, the conjugates of (23) in S_3 are

(12), (13), (23) _____

(d) Compute the conjugates of (123) in S_3:

(123) $I(123)I^{-1} =$ _____

(12)(123)(12), (132)	$(12)(123)(12)^{-1} =$ _____	= _____
(13)(123)(13), (132)	$(13)(123)(13)^{-1} =$ _____	= _____
(23)(123)(23), (132)	$(23)(123)(23)^{-1} =$ _____	= _____
(123)(123)(132), (123)	$(123)(123)(123)^{-1} =$ _____	= _____
(132)(123)(123), (123)	$(132)(123)(132)^{-1} =$ _____	= _____

Therefore, the conjugates of (123) in S_3 are

(123), (132) _____

(e) Compute the conjugates of (132) in S_3:

(132) $I(132)I^{-1} =$ _____

(12)(132)(12), (123) $(12)(132)(12)^{-1} =$ _____ = _____

242 Groups

(13)(132)(13), (123)

(23)(132)(23), (123)

(123)(132)(132), (132)

(132)(132)(123), (132)

(123), (132)

g

$xx^{-1}g, eg, g$

reflexive

symmetric, transitive

gCg

$g = xgx^{-1}$

g

bCa

$(13)(132)(13)^{-1} = $ _____ $= $ _____

$(23)(132)(23)^{-1} = $ _____ $= $ _____

$(123)(132)(123)^{-1} = $ _____ $= $ _____

$(132)(132)(132)^{-1} = $ _____ $= $ _____

Therefore, the conjugates of (132) in S_3 are

_____

123. If G is an Abelian group, then the only conjugate of an element $g \in G$ is the element _____ itself. This may be seen by this computation: for any $x \in G$

$xgx^{-1} = $ _____ $= $ _____ $= $ _____

124. *Theorem.* If G is a group, the relation C is an equivalence relation on G.

Proof. By definition of equivalence relation, we must show that C is _____, _____, and _____.

(a) *C is reflexive.* To prove that C is reflexive, we must show that if g is any member of G, then _____.

By the definition of C, in order to show that gCg, we must find some $x \in G$ such that _____.

But this is easy, since $ege^{-1} = $ _____.

Therefore, by definition of C, gCg for all $g \in G$.

(b) *C is symmetric.* To show that C is symmetric, we need to demonstrate that if aCb, then _____.

But if aCb, we have by definition of C that there

Groups 243

$a = xbx^{-1}$

is some $x \in G$ such that _____.

From $a = xbx^{-1}$, we may multiply on the left by x^{-1} and obtain _____.

$x^{-1}a = bx^{-1}$

Then from $x^{-1}a = bx^{-1}$, multiplication on the right by x gives _____.

$b = x^{-1}ax$

But since $x = (x^{-1})^{-1}$, we may write $b = x^{-1}ax$ as

$b = x^{-1}a(x^{-1})^{-1}$

_____. From this we conclude that there is an element $y \in G$ such that $b = yay^{-1}$,

x^{-1}

namely $y = $ _____. By the defini-

bCa

tion of C it follows that _____, and this

symmetric

proves that C is _____.

(c) *C is transitive.* To show that C is transitive, we must show that if gCh and hCk, for g, h, k

gCk

$\in G$, then _____. By definition of C, if gCh and hCk, then there are elements x,

$g = xhx^{-1}, h = yky^{-1}$

$y \in G$ such that _____ and _____. If $h = yky^{-1}$ is substituted into the equation

$g = x(yky^{-1})x^{-1}$

$g = xhx^{-1}$ we obtain _____ or

$y^{-1}x^{-1}$

$g = (xy)k(y^{-1}x^{-1})$. But since $(xy)^{-1} = $ _____ we may write $g = (xy)k(y^{-1}x^{-1})$ as $g = $ _____.

$(xy)k(xy)^{-1}$

Hence, there is an element $z \in G$ such that

xy

$g = zkz^{-1}$, namely, $z = $ _____.

Therefore, by definition of C, it follows that gCk

transitive

and C is _____.

125. Since C is an equivalence relation on any group G, there is a partition of G associated with C.

equivalence classes

This partition of G consists of the _____

associated with C. If $g \in G$, recall that the equivalence class $\bar{g}$ is defined as follows:

$$\bar{g} = \{h \mid h \in G \text{ and } gCh\}$$

(a) Refer to Exercise 121 to compute the following equivalence classes in $C(4)$:

$\overline{a^2} =$ _____ {a^2}

$\bar{a} =$ _____ {a}

(b) Refer to Exercise 122 to compute the following equivalence classes in S_3:

$\overline{(12)} =$ _____ {(12), (13), (23)}

$\overline{(13)} =$ _____ {(12), (13), (23)}

$\overline{(23)} =$ _____ {(23), (13), (12)}

$\overline{(123)} =$ _____ {(123), (132)}

126. The elements g, h of a group G are said to *commute* if $gh = hg$.

Theorem. If G is a group and $g \in G$, then the equivalence class $\bar{g}$ (with respect to the relation C) contains exactly one element if and only if g commutes with every element of G.

Proof. We must prove two things: (a) if $\bar{g}$ contains one element, then g commutes with every element of G; (b) if g commutes with every element of G, then $\bar{g}$ contains exactly one element.

(a) The set $\bar{g}$ contains g since $ege^{-1} =$ _____. g

Also, xgx^{-1} is conjugate to $\bar{g}$ for all $x \in G$. Therefore, if $\bar{g}$ contains only one element we must have _____ = _____. g, xgx^{-1}

Groups 245

for all $x \in G$. But from $g = xgx^{-1}$ it follows that $gx =$ _____ xg _____ for all $x \in G$; i.e., g _____ commutes _____ with every element of G.

(b) We may obtain the conjugates of g by computing xgx^{-1} for all $x \in G$. If g commutes with every element of G, then we have $xgx^{-1} =$ _____ gxx^{-1}, ge, g _____ = _____ = _____. Therefore, every conjugate of g is _____ g _____ so that the class $\bar{g}$ contains exactly _____ one _____ element.

If G is a group, the normalizer of an element $g \in G$ and the center of G may be defined in terms of the word "commute." *The normalizer $N(g)$ of an element $g \in G$ is the set of all elements in G which commute with g. The center of G is the set of all elements of G which commute with every element of G.*

127. Theorem. If G is a group and $g \in G$, then $N(g)$ is a subgroup of G.

Proof. (a) Since $N(g)$ is the set of all elements of G which commute with g, and since $ge = eg$, it follows that $e \in$ _____ $N(g)$ _____

(b) Suppose $x, y \in N(g)$. Then by definition of $N(g)$ we have $gx =$ _____ xg _____ and $gy =$ _____ yg _____ Therefore, we have $g(xy) = (gx)y$
= _____ $(xg)y$ _____

since $gx = xg$

$ = x(gy)$

$ = x(yg)$

since $gy = yg$

$ = (xy)g$

This shows that xy _____ commutes _____ with g and therefore $xy \in$ _____ $N(g)$ _____.

(c) If $x \in N(g)$, then $gx =$ _____ xg _____ by definition of _____ $N(g)$ _____. But if $gx = xg$, it follows that $gx^{-1} =$ _____ $x^{-1}g$ _____ so that x^{-1} _____ commutes _____ with g. This implies, by definition of $N(g)$, that $x^{-1} \in$ _____ $N(g)$ _____.

This completes the proof that $N(g)$ is a subgroup of G for all $g \in G$.

128. In the group S_3, compute $N((12))$. To do this we must find all those elements of S_3 which commute with (12). Two elements which obviously commute with (12) are _____ I _____ and _____ (12) _____. Hence, $I \in$ _____ $N((12))$ _____ and $(12) \in$ _____ $N((12))$ _____. To determine if $(13) \in N((12))$ we compute $(13)(12)$ and $(12)(13)$:

$(13)(12) =$ _____ (132) _____ and $(12)(13) =$ _____ (123) _____

Since $(132) \neq (123)$, it follows that $(13)(12) \neq (12)(13)$ so that $(13) \notin$ _____ $N((12))$ _____.

Next, compute $(12)(23)$ and $(23)(12)$:

$(12)(23) =$ _____ (132) _____ and $(23)(12) =$ _____ (123) _____

Groups 247

(23)(12)   Therefore, (12)(23) ≠ _____ so that
N((12))   (23) ∉ _____. Continuing with the other
            elements of S_3, we have the following:
(13), (23)   (12)(123) = _____ and (123)(12) = _____
(23), (13)   (12)(132) = _____ and (132)(12) = _____
            From these computations it follows that
N((12))   (12)(123) ≠ (123)(12) so that (123) ∉ _____.
(132)(12)   (12)(132) ≠ _____ so that (132) ∉
N((12))   _____. We then conclude that
{I, (12)}   N((12)) = _____.

129. In the group S_3, compute N((123)). For this we
     make the following computations:
(23), (13)   (123)(12) = _____ and (12)(123) = _____
(12), (23)   (123)(13) = _____ and (13)(123) = _____
(13), (12)   (123)(23) = _____ and (23)(123) = _____
I, I   (123)(132) = _____ and (132)(123) = _____
             From these computations we see that the only
             element of S_3 which commutes with (123), other
(132)   than I and (123), is _____. Therefore,
{I, (123), (132)} N((123)) = _____.

130. **Theorem.** Let G be a group, $g \in G$ and $h \in G$.
     Then $\{y \mid y = kgk^{-1}$ for $k \in hN(g)\}$ contains exactly one element.
     Proof. We show that
     $$H = \{y \mid y = kgk^{-1} \text{ for } k \in hN(g)\}$$
     contains hgh^{-1} as its only element.

248 Groups

$hN(g)$	$hgh^{-1} \in H$ since $h = he \in$ _____.
kgk^{-1}	Also, if $y \in H$, then $y =$ _____ by
$hN(g)$	definition of H and $k \in$ _____.
ht	But if $k \in hN(g)$, $k =$ _____ where
	$t \in N(g)$. Then we have $y = kgk^{-1}$
	$\qquad =$ _____
$(ht)g(ht)^{-1}$	substituting for k
	$\qquad =$ _____
$(ht)g(t^{-1}h^{-1})$	since $(ht)^{-1} = t^{-1}h^{-1}$
	$\qquad =$ _____
$h(tgt^{-1})h^{-1}$	
$h(gtt^{-1})h^{-1}$	$\qquad =$ _____
	since t commutes with g
$h(ge)h^{-1}$	$\qquad =$ _____
hgh^{-1}	$\qquad =$ _____

Therefore, hgh^{-1} is the only element of the set H.

131. Theorem. Let G be a group, $g \in G$, $h_1 \in G$ and $h_2 \in G$. Then if $h_1 N(g) \neq h_2 N(g)$, $k_1 g k_1^{-1} \neq k_2 g k_2^{-1}$ where $k_1 \in h_1 N(g)$, $k_2 \in h_2 N(g)$.

Proof. Suppose that $k_1 g k_1^{-1} = k_2 g k_2^{-1}$ where

	$k_1 = h_1 t_1$ for $t_1 \in$ _____ and
$N(g)$	$k_2 = h_2 t_2$ for $t_2 \in$ _____. Then
$N(g)$	we have, by substituting for k_1 and k_2, the following:

$$k_1 g k_1^{-1} = k_2 g k_2^{-1}$$

$(h_1 t_1)g(h_1 t_1)^{-1},\ (h_2 t_2)g(h_2 t_2)^{-1}$	_____ = _____
$(h_1 t_1)g(t_1^{-1}h_1^{-1}),\ (h_2 t_2)g(t_2^{-1}h_2^{-1})$	_____ = _____
$h_1(t_1 g t_1^{-1})h_1^{-1},\ h_2(t_2 g t_2^{-1})h_2^{-1}$	_____ = _____

Groups 249

$h_1(gt_1t_1^{-1})h_1^{-1}, h_2(gt_2t_2^{-1})h_2^{-1}$ = _____

since t_1, t_2 both commute with g

$h_1(ge)h_1^{-1}, h_2(ge)h_2^{-1}$ = _____

$h_1 g h_1^{-1}, h_2 g h_2^{-1}$ = _____

From the equation $h_1 g h_1^{-1} = h_2 g h_2^{-1}$ we obtain

$g(h_2^{-1} h_1)$ $(h_2^{-1} h_1)g = $ _____ and

$h_2^{-1} h_1$ this implies that _____ $\in N(g)$.

But if $h_2^{-1} h_1 \in N(g)$, we may conclude that

$h_2 N(g)$ $h_1 N(g) = $ _____. Therefore, if $h_1 N(g) \neq h_2 N(g)$, then $k_1 g k_1^{-1} \neq k_2 g k_2^{-1}$ where $k_1 \in h_1 N(g)$ and $k_2 \in h_2 N(g)$.

132. Theorem. Let G be a group of order n and let $g \in G$. If $\bar{g}$ is the equivalence class containing g relative to the equivalence relation C and if the number of elements in $\bar{g}$ is m, then $m = [G : N(g)]$, i.e., the number of elements in $\bar{g}$ is the index of the normalizer, $N(g)$, in G.

Proof. Since m is the index of $N(g)$ in G, there are m cosets of $N(g)$ in G. Let $h_1 N(g), h_2 N(g), h_3 N(g), \ldots, h_m N(g)$ be these m cosets. Then we have

$G = h_1 N(g) \cup h_2 N(g) \cup h_3 N(g) \cup \ldots \cup h_m N(g)$.

To construct the set $\bar{g}$ we may compute xgx^{-1} for all $x \in G$. But because of the above equation we have the following:

$\bar{g} = \{xgx^{-1} \mid x \in G\} = \{xgx^{-1} \mid x \in h_1 N(g)\} \cup \{xgx^{-1} \mid x \in h_2 N(g)\} \cup \ldots \cup \{xgx^{-1} \mid x \in h_m N(g)\}$.

According to the theorem of Exercise 130, each of the sets $\{xgx^{-1} \mid x \in h_i N(g)\}$ contains exactly _____one_____ element. Also, by the theorem of Exercise 131, none of the sets $\{xgx^{-1} \mid x \in h_i N(g)\}$ have common elements. Therefore, the total number of elements of $\bar{g}$ is the same as the total number of the sets __$\{xgx^{-1} \mid x \in h_i N(g)\}$__.
Since the total number of these sets is __m__.
we conclude that the number of elements of g is

__m__.

133. Let G be a group of order n. For each element $g \in G$ we may form the equivalence class $\bar{g}$ with respect to the equivalence relation C. The set of all these equivalence classes constitute a _____partition_____ of G. We may therefore write

$$G = \bar{g}_1 \cup \bar{g}_2 \cup \bar{g}_3 \cup \ldots \cup \bar{g}_s$$

where s is the number of equivalence classes. If we let m_i be the number of elements in the class $\bar{g}_i$, we may write

$n = $ __$m_1 + m_2 + m_3 + \ldots + m_s$__.

By the theorem of Exercise 132, each of the numbers m_i is __the index of $N(g_i)$__ in G. We may then conclude that $m_i \mid$ __n__.

134. Let G be a group of order p^k where p is a prime and k is an integer such that $k \geq 1$. If Z is the center of G, prove that Z has order greater than one and is, therefore, a multiple of p. (Hint: Use Exercises 126 and 133.)

135. Prove that if G is a group of order p^2, where p is a prime number, then G is Abelian.

136. (a) Determine the center of S_3.
(b) Determine the center of A_4.
(c) Determine the center of the quoternion group (see Exercise 117).

Appendix

Development of the Natural Numbers and Integers

The reader will find here a development of the natural numbers along the lines originated by the Italian mathematician Peano. Almost every high school graduate is familiar with the axiomatic development of Euclidean geometry. Relatively few, however, are acquainted with the axiomatic development of the number system. It will be shown here how one may begin with a small number of assumptions (axioms or postulates) and prove from these assumptions the familiar properties of the natural numbers and the integers. This material will thus serve two purposes: to acquaint the reader with the logical foundations of our number system and, at the same time, to reinforce certain facts about numbers with which he is already familiar.

A. The Natural Numbers N

Listed on p. 254 are the axioms which we will assume about N. No information about N will be used except that contained in our axioms and that which can be proved from these axioms. The reader should be very careful that he does not use

any of his previous knowledge of the natural numbers, but accepts as fact only that which can be obtained by use of the axioms.

One question before us is, "What is the set N of natural numbers?" At this point, our only answer is that the set N of natural numbers is anything for which the following axioms are true. As our development proceeds, this answer will be refined by each new result that we are able to prove.

The Peano Axioms for the Natural Numbers N

Axiom 1. The set N contains at least two elements, one of which is denoted by "1".

Axiom 2. There is a mapping $\alpha: N \longrightarrow N \cap \{1\}$ such that α is 1–1 and onto.

Axiom 3. (Induction Axiom) Suppose $S \subseteq N$ and that the following are true:
  (a) $1 \in S$;
  (b) if $x \in S$ and $\alpha: x \longrightarrow y$, then $y \in S$.
Then $S = N$.

It should be pointed out that the three axioms above make no mention of the familiar addition and multiplication of natural numbers. Our axioms are sufficient for us to *define* these binary operations in terms of the mapping α of Axiom 2. We may think of α as the mapping which gives us the "next" natural number after x if $\alpha: x \longrightarrow y$. This interpretation will be justified in equation 1.A.

Theorem A-1. Let $P(n)$ be a statement for each $n \in N$ and suppose the following are true:
  (a) $P(1)$ is true;
  (b) if $P(x)$ is true and if $\alpha: x \longrightarrow y$, then $P(y)$ is true.
Then $P(n)$ is true for all $n \in N$.

Proof. Define S as follows:
$$S = \{n \mid n \in N \text{ and } P(n) \text{ is true}\}$$
Then $S \subseteq N$ by definition of S. Since $P(1)$ is true, it follows that $1 \in S$. Now suppose that $x \in S$ and that $\alpha: x \longrightarrow y$. By definition of S, this implies that $P(x)$ is true. Hence, by hypothesis (b), it must be concluded that $P(y)$ is true and consequently that $y \in S$. This shows that the set S satisfies both (a) and (b) of Axiom 3. Therefore, by Axiom 3, $S = N$, so that $P(n)$ is true for all $n \in N$ which completes the proof.

Some of the following definitions will be given *inductively**. This means that a concept C will be defined as follows:
  (a) C will be defined for 1;

---

*For further discussion of inductive definition, the reader is referred to *Mathematical Induction* by B.K. Youse (Englewood Cliffs, N.J.: Prentice-Hall, Inc., 1964).

(b) if C is defined for y and if $\alpha: y \longrightarrow u$, then C will be defined for u.

If S is the set of natural numbers n for which the concept C is defined, it follows immediately from Axiom 3 that $S = N$ or that C is defined for all natural numbers.

DEFINITION (addition). The binary operation $+$ on N is defined as follows: if $x \in N$ and $y \in N$, then
- (a) $+ : (x, 1) \longrightarrow z$ if $\alpha: x \longrightarrow z$;
- (b) if $+ : (x, y) \longrightarrow w$, $\alpha: y \longrightarrow u$ and $\alpha: w \longrightarrow y$, then $+ : (x, u) \longrightarrow v$.

This definition of addition may be stated in the alternate notation for binary operations:
- (a) $x + 1 = z$ if $\alpha: x \longrightarrow z$;
- (b) if $x + y = w$, $\alpha: y \longrightarrow u$, and $\alpha: w \longrightarrow v$, then $x + u = v$.

Observe that $x + y$ is defined if $y = 1$ by part (a), and that if $x + y$ is defined and $\alpha: y \longrightarrow u$, then $x + u$ is defined. It then follows that $x + y$ is defined for the particular number x and *all* natural numbers y. Therefore, $x + y$ is defined for all natural numbers x, y.

The definition of addition purports to be that of a binary operation. We must actually prove that this is the case. Since we have just seen that $x + y$ is defined for all natural number pairs (x, y), it only remains to show that $+$ is a mapping. But, this is immediate, because the definition is given in terms of α and α is a 1–1 mapping.

From part (a) of the definition of addition, the mapping α can be expressed as follows:

$$\alpha: x \longrightarrow x + 1 \quad \text{for all } x \in N \tag{1.A}$$

Then using this expression for α, part (b) may be stated:

$$\text{if } x + y = w, \text{ then } x + (y + 1) = w + 1 \tag{2.A}$$

If we substitute $x + y$ for w in the second part of (2.A), we obtain that for all $x, y \in N$

$$(x + y) + 1 = x + (y + 1) \tag{3.A}$$

From now on we will use (1.A) and (2.A) instead of the original statements. Theorem A-1 may now be stated as follows:

Theorem A-1 (*restatement*). Let $P(n)$ be a statement for each $n \in N$, and suppose the following are true:
- (a) $P(1)$ is true;
- (b) if $P(x)$ is true, then $P(x + 1)$ is true.

Then $P(n)$ is true for all $n \in N$.

We are now ready to prove some of the elementary properties of addition for N.

Theorem A-2. (*The Associative Law of Addition*). If $x, y, z \in N$, then

$$(x + y) + z = x + (y + z)$$

Proof. For the particular numbers x, y we let $P(z)$ be the statement
$$(x + y) + z = x + (y + z)$$
Now we wish to show that $P(z)$ satisfies both (a) and (b) of Theorem A.1. $P(1)$ is the statement
$$(x + y) + 1 = x + (y + 1)$$
and we wish to show that this is true. But this is just the statement (3.A), and thus $P(1)$ is true by definition.

Suppose next that the statement $P(z)$ is true, i.e.,
$$(x + y) + z = x + (y + z)$$
is true. We wish to show that this implies that $P(z + 1)$ is true. We have

$$\begin{aligned}
(x + y) + (z + 1) &= ((x + y) + z) + 1 & \text{by (3.A)} \\
&= (x + (y + z)) + 1 & \text{assuming } P(z) \text{ is true} \\
&= x + ((y + z) + 1) & \text{by (3.A)} \\
&= x + (y + (z + 1)) & \text{by (3.A)}
\end{aligned}$$

Therefore, if $P(z)$ is true, it follows that
$$(x + y) + (z + 1) = x + (y + (z + 1))$$
is true or that $P(z + 1)$ is true. Hence, we conclude from Theorem A-1 that $P(z)$ is true for all $z \in N$, and this completes the proof.

Theorem A-3. If $x \in N$, then
$$x + 1 = 1 + x$$

Proof. Let $P(x)$ be the statement
$$x + 1 = 1 + x$$
$P(1)$ is obviously true. If $P(x)$ is true, we have

$$\begin{aligned}
(x + 1) + 1 &= (1 + x) + 1 & \text{since } P(x) \text{ is true} \\
&= 1 + (x + 1) & \text{by Theorem A-2}
\end{aligned}$$

Hence, if $P(x)$ is true, $P(x + 1)$ is true. Therefore, $x + 1 = 1 + x$ is true for all $x \in N$ by Theorem A-1.

Theorem A-4 (*The Commutative Law of Addition*). If $x, y \in N$, then
$$x + y = y + x$$

Proof. Let $P(y)$ be the statement
$$x + y = y + x$$
$P(1)$ is true by Theorem A-3. If $P(y)$ is true, we have

$$\begin{aligned}
x + (y + 1) &= (x + y) + 1 & \text{by Theorem A-2} \\
&= (y + x) + 1 & \text{by } P(y)
\end{aligned}$$

$$\begin{aligned} &= 1 + (y + x) & \text{by Theorem A-3} \\ &= (1 + y) + x & \text{by Theorem A-2} \\ &= (y + 1) + x & \text{by Theorem A-3} \end{aligned}$$

Thus if $P(y)$ is true, $P(y + 1)$ is true. The proof is then complete by Theorem A-1.

We now give the definition of the *product*, $x \cdot y$, for elements $x, y \in N$. This definition is also inductive and the reader will have no defficulty in showing that it gives meaning to $x \cdot y$ for all $x, y \in N$.

DEFINITION (multiplication). The binary operation $\cdot$ on N is defined as follows: if $x, y \in N$, then
(a) $1 \cdot y = y$;
(b) if $x \cdot y$ is defined, then $(x + 1) \cdot y = x \cdot y + y$.

Theorem A-5 (*The Distributive Law*). If $x, y, z \in N$ then
$$x \cdot (y + z) = x \cdot y + x \cdot z$$

Proof. Let $P(x)$ be the statement
$$x \cdot (y + z) = x \cdot y + x \cdot z$$
and apply Theorem A-1. The statement $P(1)$ is
$$1 \cdot (y + z) = 1 \cdot y + 1 \cdot z$$
Applying part (a) of the definition of multiplication, we have
$$1 \cdot (y + z) = y + z$$
and
$$1 \cdot y + 1 \cdot z = y + z$$
Hence, $P(1)$ is true. Now suppose that $P(x)$ is true. Then we wish to show that $P(x + 1)$ is true.

$$\begin{aligned} (x + 1) \cdot (y + z) &= x \cdot (y + z) + (y + z) & \text{by part (b) of definition of } \cdot \\ &= (x \cdot y + x \cdot z) + (y + z) & \text{by } P(x) \\ &= ((x \cdot y + x \cdot z) + y) + z & \text{by Theorem A-2} \\ &= (x \cdot y + (x \cdot z + y)) + z & \text{by Theorem A-2} \\ &= (x \cdot y + (y + x \cdot z)) + z & \text{by Theorem A-4} \\ &= ((x \cdot y + y) + x \cdot z) + z & \text{by Theorem A-2} \\ &= ((x + 1) \cdot y + x \cdot z) + z & \text{by (a) of definition of } \cdot \\ &= (x + 1) \cdot y + (x \cdot z + z) & \text{by Theorem A-2} \\ &= (x + 1) \cdot y + (x + 1) \cdot z & \text{by (a) of definition of } \cdot \end{aligned}$$

Therefore, if $P(x)$ is true, it follows that
$$(x + 1) \cdot (y + z) = (x + 1) \cdot y + (x + 1) \cdot z$$

is true, i.e., $P(x + 1)$ is true. Thus by Theorem A-1, $P(x)$ is true for all $x \in N$, and this proves the theorem.

Theorem A-6 (*The Commutative Law for Multiplication*). If $x, y \in N$, then
$$x \cdot y = y \cdot x$$
Proof. Let $P(x)$ be the statement $x \cdot y = y \cdot x$. Then $P(1)$ is
$$1 \cdot y = y \cdot 1$$
Since $1 \cdot y = y$ by part (a) of the definition of multiplication, we must show that $y \cdot 1 = y$ for all $y \in N$. To do this, let $Q(y)$ be the statement $y \cdot 1 = y$. Then $Q(1)$ is true by part (a) of the definition of multiplication. If $Q(y)$ is true, we have

$$\begin{aligned}(y + 1) \cdot 1 &= y \cdot 1 + 1 \cdot 1 &&\text{by (b) of definition of } \cdot \\ &= y + 1 \cdot 1 &&\text{since } Q(y) \text{ is true} \\ &= y + 1 &&\text{since } Q(1) \text{ is true}\end{aligned}$$

Therefore, if $Q(y)$ is true, $Q(y + 1)$ is true, and it follows from Theorem A-1 that $y \cdot 1 = y$ for all $y \in N$. We may then conclude that
$$1 \cdot y = y \cdot 1$$
is true for all $y \in N$ and consequently that $P(1)$ is true. Now suppose that $P(x)$ is true. Then

$$\begin{aligned}(x + 1) \cdot y &= x \cdot y + y &&\text{by (b) of definition of } \cdot \\ &= y \cdot x + y &&\text{by } P(x) \\ &= y \cdot x + y \cdot 1 &&y \cdot 1 = y\text{—just proved} \\ &= y \cdot (x + 1) &&\text{by Theorem A-5}\end{aligned}$$

Hence, if $P(x)$ is true, $P(x + 1)$ is true. By Theorem A-1, the proof is complete.

Theorem A-7 (*The Associative Law of Multiplication*). If $x, y, z \in N$, then
$$x \cdot (y \cdot z) = (x \cdot y) \cdot z$$
Proof. This may be proved by Theorem A.1, and is left to the reader.

B. The Order Relation for N

We are now in a position to define the order relations "less than" and "greater than" for N.

DEFINITION (Less than). If $x, y \in N$, we say that $x < y$ if and only if there is $a \in N$ such that $x + a = y$. We say that
$$y > x \text{ if and only if } x < y$$
(The statement $x < y$ is read "x is less than y"; $y > x$ is read "y is greater than x.")

Appendix 259

Theorem B-1. The relations "less than" and "greater than" are both transitive. Neither of these relations is reflexive nor symmetric.

Proof. The parts of this theorem about $<$ will be proved and the parts concerning $>$ will be left for the reader to prove. To show that $<$ is transitive, it must be shown that for $x, y, z \in N$

$$x < y \text{ and } y < z \text{ implies } x < z$$

By definition of $<$, if $x < y$ and $y < z$, there are elements $a, b \in N$ such that $x + a = y$ and $y + b = z$. Substituting for y in the second equation gives

$$(x + a) + b = z$$

or

$$x + (a + b) = z$$

Since $a + b \in N$, it follows by the definition of $<$ that $x < z$. Hence, $<$ is transitive.

To prove that $<$ is not reflexive, we must show it is not true that $x < x$ for all $x \in N$. We must show that there is at least one element $a \in N$ such that $a < a$ is not true. We may easily show it is false that $1 < 1$. To accomplish this we only have to show that $1 + b \neq 1$ for all $b \in N$. This will demonstrate that there is no $b \in N$ such that $1 + b = 1$, and hence it is false (by definition of $<$) that $1 < 1$. By Theorem A-3, $1 + b = b + 1$ and by Axiom 2, $\alpha: b \rightarrow b + 1$. If $1 + b = 1$, we would have $\alpha : b \rightarrow 1$. But by Axiom 2, $\alpha: N \rightarrow N \cap \{1\}$. Therefore, we cannot have $\alpha: b \rightarrow 1$ and consequently, $1 + b \neq 1$ for all $b \in N$. It then follows that $<$ is not reflexive. (Actually, it is true that $x < x$ for all $x \in N$; the reader should prove this.)

To show that $<$ is not symmetric, it must be shown it is not true that if $x < y$, then $y < x$ for all $x, y \in N$. By definition of $<$ it is true that $1 < 1 + 1$, and we will show that $1 + 1 \not< 1$; this will complete the proof. If it is true that $1 + 1 < 1$, there is $a \in N$ such that $(1 + 1) + a = 1$ or by Theorem A-2 and Theorem A-3, $(1 + a) + 1 = 1$. By Axiom 2, $\alpha: 1 + a \rightarrow (1 + a) + 1$, and if $(1 + a) + 1 = 1$, then $\alpha: 1 + a \rightarrow 1$, contrary to the fact that $\alpha: N \rightarrow N \cap \{1\}$.

Lemma 1. Let $a \in N$. Then $x + a \neq x$ for all $x \in N$ or $x \not< x$ for all $x \in N$.

Proof. Let $P(x)$ be the statement $x + a \neq x$. The truth of $P(1)$ was established in the proof of Theorem B-1. Suppose that $P(x)$ is true, or that $x + a \neq x$. By Axiom 2, $\alpha: x \rightarrow x + 1$ and $\alpha: x + a \rightarrow (x + a) + 1$. But by Theorem A-2 and Theorem A-3, $(x + a) + 1 = (x + 1) + a$ so that $\alpha: x + a \rightarrow (x + 1) + a$. Therefore, if $x + a \neq x$, then $(x + 1) + a \neq x + 1$ since α is a 1-1 mapping. Thus the truth of $P(x)$ implies the truth of $P(x + 1)$. By Theorem A-1, the lemma is proved.

Lemma 2. Let $x, y \in N$. Then $x < y$ implies $x + 1 < y$ or $x + 1 = y$.

Proof. If $x < y$, there is some $a \in N$ such that $x + a = y$ by definition of $<$. If $a = 1$, $x + 1 = y$ and the result is proved. If $a \neq 1$, then $a \in N \cap \{1\}$, and by Axiom 2, there is a $b \in N$ such that $\alpha: b \rightarrow a$ or $a = b + 1$. Then

$$y = x + a = x + (b + 1)$$
$$= x + (1 + b) \quad \text{By Theorem A-3}$$
$$= (x + 1) + b \quad \text{By Theorem A-2}$$

Therefore, $(x + 1) + b = y$ and by definition of $<$, $x + 1 < y$. The lemma is proved.

Theorem B-2. (*The Trichotomy Property of N*). Let $x, y \in N$. Then
   (a) At least one of the following statements is true:

   (i) $x < y$ (ii) $x = y$ (iii) $y < x$

   (b) Not more than one of the following statements is true:

   (i) $x < y$ (ii) $x = y$ (iii) $y < x$

Proof. (a) To prove this part of the theorem, we let $P(x)$ be the statement that at least one of the parts,

   (i) $x < y$ (i) $x = y$ (iii) $y < x$

is true. Then $P(1)$ asserts that one of the following is true: $1 < y$, $1 = y$ or $y < 1$. If $y = 1$, we have finished. If $y \neq 1$, then as in the proof of Lemma 2, there is $b \in N$ such that $y = b + 1$. It follows that $1 < b + 1$ by definition of $<$ and hence that $1 < y$. Therefore, $P(1)$ is true. Now suppose that $P(x)$ is true or that at least one of (i), (ii), (iii) is true. If $x < y$, then by Lemma 2, $x + 1 < y$ or $x + 1 = y$. If $x = y$, then $x < x + 1$ by definition of $<$ or $y < x + 1$. If $y < x$, there is an element $b \in N$ such that $y + b = x$. Then $(y + b) + 1 = x + 1$ and by Theorem A-2, $y + (b + 1) = x + 1$. By definition of $<$ it follows that $y < x + 1$. Thus, in all cases it has been shown that if $P(x)$ is true, then $P(x + 1)$ is true. This proves part (a) by Theorem A-1.

   (b) If $x < y$ and $x = y$, then $x < x$, which contradicts Lemma 1. If $y < x$ and $x = y$, then Lemma 1 is again contradicted. If $x < y$ and $y < x$, then since $<$ is transitive by Theorem B-1, $x < x$. This is another contradiction of Lemma 1, and shows that it is not possible for two parts of (i), (ii), (iii) to hold, which completes the proof.

Theorem B-3. Let $a \in N$. Then for all $x, y \in N$,
   (a) $x < y$ implies $x + a < y + a$ and $x \cdot a < y \cdot a$;
   (b) $x + a < y + a$ implies $x < y$;
   (c) $x = y$ implies $x + a = y + a$ and $x \cdot a = x \cdot a$;
   (d) $x + a = y + a$ implies $x = y$;
   (e) $x \cdot a = y \cdot a$ implies $x = y$;
   (f) $x \cdot a < y \cdot a$ implies $x < y$.

Proof. (a) If $x < y$, then by definition of $<$ there is some $b \in N$ such that $x + b = y$. Then $(x + b) + a = y + a$, or by Theorem A-2 and Theorem A-4, $(x + a) + b = y + a$. By definition of $<$ it follows that $x + a < y + a$. Also from

$x + b = y$, we have $(x + b) \cdot a = y \cdot a$ or, by Theorem A-5 and Theorem A-6, $x \cdot a + b \cdot a = y \cdot a$. Since $b \cdot a \in N$, we conclude that $x \cdot a < y \cdot a$ by definition of $<$.

(b) Suppose that $x + a < y + a$. By Theorem B-2, we must have $x < y$, $x = y$ or $y < x$. If $y < x$, then $y + a < x + a$ by part (a). But we would then have both $x + a < y + a$ and $y + a < x + a$, contrary to Theorem B-2. Also, if $x = y$, we have $x + a = y + a$ by definition of equality and the fact that $+$ is a binary operation. This would give $x + a < y + a$ and $x + a = y + a$—again contradicting Theorem B-2. Therefore, if $x + a < y + a$, $x < y$.

(c) This follows from the fact that both $+$ and $\cdot$ are binary operations on N, and from the definition of $=$.

(d) This is proved in a manner similar to the proof of part (b).

(e) Left to the reader.

(f) Left to the reader.

DEFINITION. Let $X \subseteq N$. Suppose $a \in X$ and that the following is true:

$$a < x \text{ or } a = x \text{ for all } x \in X$$

Then a is called the *smallest element* of X.

The reader will recall that the Well-Ordering Principle for N was assumed to hold in Chapter 3. We are now able to prove this important result.

Theorem B-4. (*The Well-Ordering Principle for N*). If $X \subseteq N$ and $X \neq \emptyset$, then X contains a smallest element.

Proof. Suppose that $1 \in X$. Then 1 is the smallest element of X. For if $s \in X$ and $x \neq 1$, then $x \in N \cap \{1\}$, and by Axiom 2, there is some $y \in N$ such that $\alpha: y \to x$. But $\alpha: y \to y + 1$ and α is 1-1. Therefore $x = y + 1$ and this implies, by definition of $<$, that $1 < x$. Now suppose that $1 \notin X$ and that X has no smallest element. Define a set Y as follows:

$$Y = \{y \mid y \in N, y < x \text{ for all } x \in X\}$$

Since $1 \notin X$, we have, by the preceding argument, that $1 \in Y$. Observe that $X \cap Y = \emptyset$. Now suppose that $y \in Y$. Then by definition, $y < x$ for all $x \in X$. By Lemma 2, if $y < x$, then $y + 1 < x$ or $y + 1 = x$. If $y + 1 = x$ for some $x \in X$, then $y + 1$ is the smallest element of X. For if $x_1 \in X$ and $y + 1 \not< x_1$ and $y + 1 \neq x_1$, then $x_1 < y + 1$ by Theorem B.2. But if $x_1 < y + 1$, we have by Lemma 2 that $x_1 + 1 < y + 1$ or $x_1 + 1 = y + 1$. Then by Theorem B-3, $x_1 < y$ or $x_1 = y$. In either case we get a contradiction, since $X \cap Y = \emptyset$. We must then conclude that $y + 1 \neq x$ for all $x \in X$ and that $y + 1 < x$ for all $x \in X$, which shows that if $y \in Y$, then $y + 1 \in Y$. Since we have already shown that $1 \in Y$, it follows from Axiom 3 that $Y = N$. By hypothesis, $X \neq \emptyset$. Hence, there is an element $a \in X$ and by definition of Y, $a \notin Y$. This implies that $Y \neq N$—a contradiction. Thus our assumption that X has no smallest element has led to a contradiction, and this proves the theorem.

We have now proved many of the properties of the natural numbers with which we are familiar. It should be pointed out that the symbols "2", "3", "4", etc., are *defined* in the usual way; 2 is defined to mean $1 + 1$, 3 means $2 + 1$, 4 means $3 + 1$, etc. Before turning our attention to the integers, we will summarize our development of the natural numbers.

Summary of Properties of the Natural Numbers

The natural numbers comprise a set N with two binary operations, $+$, $\cdot$, and an order relation, $<$, for which the following properties hold:

1. *Commutative Laws:* For all $x, y \in N$
$$x + y = y + x \text{ and } x \cdot y = y \cdot x$$

2. *Associative Laws:* For all $x, y, z \in N$
$$(x + y) + z = x + (y + z) \text{ and } (x \cdot y) \cdot z = x \cdot (y \cdot z)$$

3. *Distributive Law:* For all $x, y, z \in N$
$$x \cdot (y + z) = x \cdot y + x \cdot z$$

4. *Trichotomy Law:* For all $x, y \in N$, one and only one of the following statements is true:
   (i) $x < y$ (ii) $x = y$ (iii) $y < x$

5. *Cancellation Laws:* For all $x, y, z \in N$
   (a) $x + z = y + z$ implies $x = y$;
   (b) $x \cdot z = y \cdot z$ implies $x = y$;
   (c) $x + z < y + z$ implies $x < y$;
   (d) $x \cdot z < y \cdot z$ implies $x < y$.

6. *Well-Ordering Principle:* If $X \subseteq N$ and $X \neq \emptyset$, then X contains a smallest element.

C. The Integers Z

Now that the natural numbers have been developed, the integers may be defined and their elementary properties set forth. Our knowledge of the integers may be based entirely on the natural numbers by introducing integers as equivalence classes of natural numbers.

DEFINITION. The relation I is defined on the set $N \times N$ as follows: $(a, b) \, I \, (c, d)$ if and only if $a + d = c + b$ for all $a, b, c, d \in N$.

Theorem C-1. The relation I is an equivalence relation on $N \times N$.

Proof. (a) (Reflexive) It must be shown that $(a, a) I (a, a)$ for all $a \in N$. But from the definition of I, this only requires that $a + a = a + a$, which is true. Hence, I is reflexive.

(b) (Symmetric) We must show that if $(a, b) I (c, d)$, then $(c, d) I (a, b)$. By definition of I, if $(a, b) I (c, d)$, then $a + d = c + b$ or $c + b = a + d$. Then by definition of I again, if $c + b = a + d$, $(c, d) I (a, b)$. Therefore, I is symmetric.

(c) (Transitive) To show that I is transitive, suppose that $(a, b) I (c, d)$ and $(c, d) I (e, f)$. From the definition of I, this implies that $a + d = c + b$ and $c + f = e + d$. Then we have

$$
\begin{aligned}
(a + f) + c &= a + (f + c) && \text{associative law for } + \\
&= a + (c + f) && \text{commutative law for } + \\
&= a + (e + d) && \text{since } c + f = e + d \\
&= a + (d + e) && \text{commutative law for } + \\
&= (a + d) + e && \text{associative law for } + \\
&= (c + b) + e && \text{since } a + d = c + b \\
&= (b + c) + e && \text{commutative law for } + \\
&= e + (b + c) && \text{commutative law for } + \\
&= (e + b) + c && \text{associative law for } +
\end{aligned}
$$

Hence, we conclude that $(a + f) + c = (e + b) + c$. By the cancellation law for $+$, it then follows that $a + f = e + b$. The definition of I then implies that $(a, b) I (e, f)$. It has been shown that if $(a, b) I (c, d)$ and $(c, d) I (e, f)$, then $(a, b) I (e, f)$. This shows that I is transitive and completes the proof.

Since I is an equivalence relation on $N \times N$, there is a partition of $N \times N$ associated with I which consists of the equivalence classes $\overline{(a, b)}$ for $a, b \in N$.

DEFINITION. An equivalence class $\overline{(a, b)}$ associated with the equivalence relation I is called an *integer*. The set of integers is denoted by Z.

We remind the reader that if R is an equivalence relation and $\bar{x}, \bar{y}$ are two equivalence classes associated with R, then $\bar{x} = \bar{y}$ if and only if $x R y$. In the present case, $\overline{(a, b)} = \overline{(c, d)}$ if and only if $(a, b) I (c, d)$ or if and only if $a + d = c + b$. From this it follows that $\overline{(a, a)} = \overline{(b, b)}$ for all $a, b \in N$, since $a + b = b + a$. We then define the symbol "0" to mean the equivalence class $\overline{(x, x)}$ for any $x \in N$, or $0 = \overline{(x, x)}$.

Suppose that $a, b \in N$ and $a \neq b$. Then by the Trichotomy Law, $a < b$ or $b < a$. If $a < b$, there is an $n \in N$ such that $a + n = b$. Hence, the equivalence class $\overline{(a, b)}$ can be written as $\overline{(a, a + n)}$ and the equivalence class $\overline{(b, a)}$ can be written as $\overline{(a + n, a)}$.

DEFINITION. The symbols "0", "^+n" and "^-n" are defined as follows:
   (a) $0 = \overline{(x, x)}$ for any $x \in N$

(b) $^+n = \overline{(a+n, a)}$ for any $a \in N$
(c) $^-n = \overline{(a, a+n)}$ for any $a \in N$

It should be noted that parts (b) and (c) of the definition are justified by the fact that $\overline{(a+n, a)} = \overline{(x+n, x)}$ and $\overline{(a, a+n)} = \overline{(x, x+n)}$ for all $a, x \in N$. The integer 0 is called "zero"; the integers ^+n are called *positive* integers while the integers ^-n are called *negative* integers.

DEFINITION (For addition and multiplication on Z). Addition, $+$, and *multiplication*, $\cdot$, on Z are defined as follows: for all $\overline{(a, b)}, \overline{(c, d)} \in Z$,

$$\overline{(a, b)} + \overline{(c, d)} = \overline{(a+c, b+d)}$$
$$\overline{(a, b)} \cdot \overline{(c, d)} = \overline{(ac+bd, ad+bc)}$$

The definitions of $+$ and $\cdot$ are of binary operations on Z, but this is not obvious. For example, it is conceivable that the following situation could arise:

$$\overline{(a, b)} = \overline{(a', b')}, \overline{(c, d)} = \overline{(c', d')}$$

but

$$\overline{(a, b)} + \overline{(c, d)} \neq \overline{(a', b')} + \overline{(c', d')}$$

If this were possible, it would, of course, prevent $+$ from being a binary operation on Z. But if $\overline{(a, b)} = \overline{(a', b')}$ and $\overline{(c, d)} = \overline{(c', d')}$, it follows that

$$a + b' = b + a'$$
$$c + d' = d + c'$$
(1.C)

Now, using the definition of $+$, we have

$$\overline{(a, b)} + \overline{(c, d)} = \overline{(a+c, b+d)}$$

and

$$\overline{(a', b')} + \overline{(c', d')} = \overline{(a'+c', b'+d')}$$

From Equation (1.C)

$$(a + b') + (c + d') = (b + a') + (d + c')$$

or by using the commutative and associative laws for $+$

$$(a + c) + (b' + d') = (a' + c') + (b + d) \qquad (2.C)$$

From (2.C) it follows that

$$\overline{(a+c, b+d)} = \overline{(a'+c', b'+d')}$$

or that

$$\overline{(a, b)} + \overline{(c, d)} = \overline{(a', b')} + \overline{(c', d')}$$

This, therefore, shows that if $\overline{(a, b)} = \overline{(a', b')}$ and $\overline{(c, d)} = \overline{(c', d')}$, then

$$\overline{(a, b)} + \overline{(c, d)} = \overline{(a', b')} + \overline{(c', d')}$$

Hence, $+$ is a binary operation on Z. In a similar fashion it is possible to show that $\cdot$ is also a binary operation on Z.

Theorem C-2. *For all $u, v, w \in Z$ and $n \in N$, the following laws hold for $+$ and $\cdot$:*

1. *Commutative Laws*
$$u + v = v + u \text{ and } u \cdot v = v \cdot u$$

2. *Associative Laws*
$$(u + v) + w = u + (v + w) \text{ and } (u \cdot v) \cdot w = u \cdot (v \cdot w)$$

3. *Distributive Law*
$$u \cdot (v + w) = u \cdot v + u \cdot w$$

4. *The Identity Property of 0 for $+$*
$$u + 0 = 0 + u = u$$

5. *The Identity Property of $^+1$ for $\cdot$*
$$u \cdot {}^+1 = {}^+1 \cdot u = u$$

6. *The Inverse Property for $+$*
$$^+n + {}^-n = 0$$

7. *The Cancellation Law for $+$*
$$\text{If } u + v = w + v, \text{ then } u = w$$

Proof. By definition of integers, u, v, w are equivalence classes associated with I. Thus there are natural numbers a, b, c, d, e, f such that
$$u = \overline{(a, b)}, \quad v = \overline{(c, d)}, \quad w = \overline{(e, f)}$$

1. To prove that $u + v = v + u$, we first use the definition of $+$ to compute $u + v$ and $v + u$:

$$\begin{aligned} u + v &= \overline{(a, b) + (c, d)} \\ &= \overline{(a + c, b + d)} \quad \text{by definition of } + \\ v + u &= \overline{(c, d) + (a, b)} \\ &= \overline{(c + a, d + b)} \quad \text{by definition of } + \end{aligned}$$

Now, since $(a + c) + (d + b) = (c + a) + (b + d)$, by the commutative and associative laws of $+$ for natural numbers, it follows that $\overline{(a + c, b + d)} = \overline{(c + a, d + b)}$ and, therefore, $u + v = v + u$.

We prove that $u \cdot v = v \cdot u$ in a similar way:

$$\begin{aligned} u \cdot v &= \overline{(a, b) \cdot (c, d)} \\ &= \overline{(ac + bd, ad + bc)} \quad \text{by definition of } \cdot \\ v \cdot u &= \overline{(c, d) \cdot (a, b)} \\ &= \overline{(ca + db, cb + da)} \quad \text{by definition of } \cdot \end{aligned}$$

By the associative and commutative laws of $+$ and the commutative law of $\cdot$ for natural numbers,
$$(ac + bd) + (cb + da) = (ca + db) + (ad + bc)$$
It then follows that
$$\overline{(ac + bd, ad + bc)} = \overline{(ca + db, cb + da)}$$
and, hence $u \cdot v = v \cdot u$.

2. The proof of the associative laws is quite similar to that for the commutative laws and is left to the reader.

3. As with the commutative and associative laws for Z, the distributive law for Z depends upon the corresponding law for N, as the following computations show.
$$\begin{aligned} u \cdot (v + w) &= \overline{(a, b)} \cdot (\overline{(c, d)} + \overline{(e, f)}) \\ &= \overline{(a, b)} \cdot \overline{(c + e, d + f)} \quad \text{by definition of } + \\ &= \overline{(a \cdot (c + e) + b \cdot (d + f), a \cdot (d + f) + b \cdot (c + e))} \quad \text{by definition of } \cdot \end{aligned}$$
$$\begin{aligned} u \cdot v + u \cdot w &= \overline{(a, b)} \cdot \overline{(c, d)} + \overline{(a, b)} \cdot \overline{(e, f)} \\ &= \overline{(ac + bd, ad + bc)} + \overline{(ae + bf, af + be)} \\ &\quad \text{by definition of } \cdot \\ &= \overline{((ac + bd) + (ae + bf), (ad + bc) + (af + be))} \quad \text{by definition of } + \end{aligned}$$

Using the associative and commutative laws of $+$ for natural numbers and the distributive law for N, we easily show that
$$\begin{aligned} &(a \cdot (c + e) + b \cdot (d + f)) + ((ad + bc) + (af + be)) \\ &= ((ac + bd) + (ae + bf)) + (a(d + f) + b(c + e)) \end{aligned}$$
It then follows that
$$\begin{aligned} &\overline{(a \cdot (c + e) + b \cdot (d + f), a \cdot (d + f) + b \cdot (c + e))} \\ &= \overline{((ac + bd) + (ae + bf), (ad + bc) + (af + be))} \end{aligned}$$
and, hence, that
$$u \cdot (v + w) = u \cdot v + u \cdot w$$

4. By definition, $0 = \overline{(x, x)}$ for any $x \in N$. Then
$$\begin{aligned} u + 0 &= \overline{(a, b)} + \overline{(x, x)} \\ &= \overline{(a + x, b + x)} \quad \text{by definition of } + \end{aligned}$$
But it is clear that
$$(a + x) + b = a + (b + x)$$
by the commutative and associative laws of $+$ for N. Therefore, $\overline{(a + x, b + x)} = \overline{(a, b)}$ or $u + 0 = u$. In a similar way we may prove that $0 + u = u$.

5. By definition of 1, we have
$$1 = \overline{(x+1, x)}$$
for any $x \in N$. Thus
$$u \cdot 1 = \overline{(a,b)} \cdot \overline{(x+1, x)}$$
$$= \overline{(a \cdot (x+1) + b \cdot x, a \cdot x + b \cdot (x+1))} \quad \text{by definition of } \cdot$$

By the associative and commutative laws of $+$ for N and the distributive law for N, we have
$$(a \cdot (x+1) + b \cdot x) + b = a + (a \cdot x + b \cdot (x+1))$$
and this implies that
$$\overline{(a \cdot (x+1) + b \cdot x, a \cdot x + b \cdot (x+1))} = \overline{(a,b)} = u$$

Hence, $u \cdot 1 = u$, and in a similar way, $1 \cdot u = u$.

6. By definition of ^+n and ^-n, we have
$$^+n = \overline{(x+n, x)}, \quad ^-n = \overline{(y, y+n)}$$
for any $x, y \in N$. Consequently,
$$^+n + ^-n = \overline{(x+n, x)} + \overline{(y, y+n)}$$
$$= \overline{((x+n) + y, x + (y+n))} \quad \text{by definition of } +$$

This last equivalence class is, by definition, 0, since $(x+n) + y = x + (y+n)$.

7. Suppose that $u + v = w + v$. Then
$$\overline{(a,b)} + \overline{(c,d)} = \overline{(e,f)} + \overline{(c,d)}$$
or
$$\overline{(a+c, b+d)} = \overline{(e+c, f+d)} \quad \text{by definition of } +$$

From this it follows that
$$(a+c) + (f+d) = (e+c) + (b+d)$$

Using the associative law of $+$ for N on this last equation, we may conclude that
$$((a+c) + f) + d = ((e+c) + b) + d$$

Then by the cancellation law of $+$ for N, it follows that
$$(a+c) + f = (e+c) + b$$
or that
$$(a+f) + c = (e+b) + c$$
by the appropriate laws for N. Applying the proper cancellation law of N once more to the above equation gives
$$a + f = e + b$$

The equation $a + f = e + b$ implies that $\overline{(a,b)} = \overline{(e,f)}$ or that $u = w$.

D. The Order Relations for Z

In the previous section, the binary operations $+$ and $\cdot$ were defined on Z in terms of the corresponding binary operations on N. In a similar way, we now define the order relation "less than" on Z in terms of the relation $<$ defined on N.

DEFINITION. For integers $\overline{(a, b)}$ and $\overline{(c, d)}$ we say that $\overline{(a, b)}$ *is less than* $\overline{(c, d)}$ if and only if $a + d < c + b$.

The symbol "$<$" is also used for the "less than" relation on Z. The definition may be expressed with this symbol as follows:

$$\overline{(a, b)} < \overline{(c, d)} \text{ if and only if } a + d < c + b$$

It has been implied that $<$ is a relation on Z. Actually, this must be proved; i.e., we must show that if $\overline{(a, b)} < \overline{(c, d)}, \overline{(a, b)} = \overline{(a', b')}$ and $\overline{(c, d)} = \overline{(c', d')}$, then $\overline{(a', b')} < \overline{(c', d')}$. From $\overline{(a, b)} < \overline{(c, d)}$, $\overline{(a, b)} = \overline{(a', b')}$ and $\overline{(c, d)} = \overline{(c', d')}$, we may conclude the following:

$$a + d < c + b \tag{1.D}$$
$$a + b' = a' + b \tag{2.D}$$
$$c + d' = c' + d \tag{3.D}$$

Now consider the following:

$$\begin{aligned}
(a' + d') + (c + b) &= ((a' + d') + c) + b & \text{associative law for } + \\
&= (a' + (d' + c)) + b & \text{associative law for } + \\
&= (a' + (c + d')) + b & \text{commutative law for } + \\
&= (a' + (c' + d)) + b & \text{by 3.D} \\
&= a' + ((c' + d) + b) & \text{associative law for } + \\
&= a' + (b + (c' + d)) & \text{commutative law for } + \\
&= (a' + b) + (c' + d) & \text{associative law for } + \\
&= (a + b') + (c' + d) & \text{by 2.D} \\
&= a + ((b' + c') + d) & \text{associative law for } + \\
&= a + (d + (b' + c')) & \text{commutative law for } + \\
&= (a + d) + (b' + c') & \text{associative law for } +
\end{aligned}$$

Therefore, $(a' + d') + (c + b) = (a + d) + (b' + c')$. From 1.D and Theorem B-3 we may conclude that

$$(a + d) + (b' + c') < (c + b) + (b' + c')$$

and, hence, that
$$(a' + d') + (c + b) < (c + b) + (b' + c')$$
and
$$(a' + d') + (c + b) < (b' + c') + (c + b)$$
by the commutative law for $+$. Then again applying Theorem B-3,
$$a' + d' < b' + c'$$
or
$$a' + d' < c' + b'$$
From this it follows that $\overline{(a', b')} < \overline{(c', d')}$. This proves, then, that $<$ is indeed a relation on Z.

Theorem D-1 (*The Trichotomy Law*). If $u, v \in Z$, then one and only one of the following statements is true:
$$u < v \qquad u = v \qquad v < u$$

Proof. Let $u = \overline{(a, b)}$ and $v = \overline{(c, d)}$ for $a, b, c, d \in N$. Then the three statements of the theorem may be expressed as
$$a + d < c + b \qquad a + d = c + b \qquad c + b < a + d$$
Now by the Trichotomy Property for N, one and only one of these statements is true, and this proves the theorem.

Theorem D-2. Let $u, v, w \in Z$. Then
  (a) $u < v$ implies $u + w < v + w$;
  (b) $u + w < v + w$ implies $u < v$;
  (c) if $u < v$ and $0 < w$, then $u \cdot w < v \cdot w$;
  (d) if $u < v$ and $w < 0$, then $v \cdot w < u \cdot w$;
  (e) if $u \cdot w < v \cdot w$ and $0 < w$, then $u < v$;
  (f) if $u \cdot w < v \cdot w$ and $w < 0$, then $v < u$.

Proof. Let $u = \overline{(a, b)}, v = \overline{(c, d)}$ and $w = \overline{(e, f)}$ for $a, b, c, d, e, f \in N$.
  (a) If $u < v$, i.e., $\overline{(a, b)} < \overline{(c, d)}$, then by definition of $<$
$$a + d < c + b$$
From this it follows by Theorem B-3(a) that
$$(a + d) + (e + f) < (c + b) + (e + f) \tag{4.D}$$
Using the commutative and associative laws for $+$, 4.D may be expressed as
$$(a + e) + (d + f) < (c + e) + (b + f) \tag{5.D}$$

But 5.D implies, by definition of $<$, that
$$\overline{(a+e, b+f)} < \overline{(c+e, d+f)} \qquad (6.D)$$
Finally, since $\overline{(a+e, b+f)} = \overline{(a, b)} + \overline{(e, f)}$ and $\overline{(c+e, d+f)} = \overline{(c, d)} + \overline{(e, f)}$ 6.D is
$$\overline{(a, b)} + \overline{(e, f)} < \overline{(c, d)} + \overline{(e, f)}$$
or
$$u + w < v + w$$

(b) The steps of part (a) may be reversed to prove part (b).
(c) If $u < v$, then $a + d < c + b$. We are given here that $0 < w$ or that
$$\overline{(x, x)} < \overline{(e, f)}$$
for any $x \in N$. This implies that
$$x + f < e + x$$
or, by Theorem B-3(b) and the commutative law of $+$, that
$$f < e$$
Then there is $y \in N$ such that
$$f + y = e \qquad (7.D)$$
Since $a + d < c + b$, we may apply Theorem B.3(a) to obtain
$$(a + d) \cdot y < (c + b) \cdot y \qquad (8.D)$$
From 8.D it follows that
$$(a + d) \cdot y + (a + d) \cdot f < (c + b) \cdot y + (a + d) \cdot f \qquad (9.D)$$
by using Theorem B-3(a). From 9.D,
$$(a + d) \cdot (y + f) + (b + c) \cdot f < (b + c) \cdot f + (c + b) \cdot y + (a + d) \cdot f$$
or
$$(a + d) \cdot (f + y) + (b + c) \cdot f < (b + c) \cdot (f + y) + (a + d) \cdot f \qquad (10.D)$$
From 7.D, substitute e for $f + y$ in 10.D:
$$(a + d) \cdot e + (b + c) \cdot f < (b + c) \cdot e + (a + d) \cdot f \qquad (11.D)$$
Then, using the distributive, associative and commutative laws, we may write 11.D as
$$(a \cdot e + b \cdot f) + (c \cdot f + d \cdot e) < (c \cdot e + d \cdot f) + (a \cdot f + b \cdot e) \qquad (12.D)$$
By definition of $<$ for Z, 12.D implies that
$$\overline{(a \cdot e + b \cdot f, a \cdot f + b \cdot e)} < \overline{(c \cdot e + d \cdot f, c \cdot f + d \cdot e)}$$
or
$$\overline{(a, b)} \cdot \overline{(e, f)} < \overline{(c, d)} \cdot \overline{(e, f)}$$
Therefore, if $u < v$ and $0 < w$, then $u \cdot w < v \cdot w$.

(d) The proof of this part is similar to the proof of part (c).
(e) The steps of part (c) may be reversed to prove part (e).
(f) Part (f) is similar to (e).

Theorem D-3 (*The Cancellation Law for* $\cdot$ *in Z*). If $u, v, w \in Z$, $w \neq 0$ and if $u \cdot w = v \cdot w$, then $u = v$.

Proof. Suppose that $u \cdot w = v \cdot w$. By the Trichotomy Property of Z we know that $u < v$, $u = v$ or $v < u$. If either $u < v$ or $v < u$, parts (e) and (d) of Theorem D-2 imply (with the Trichotomy Property, again) that $u \cdot w \neq v \cdot w$. Therefore, since we can not have either $u < v$ or $v < u$, it must be that $u = v$.

Theorem D-4. Let $u, v \in Z$. Then
  (a) $u \cdot 0 = 0$;
  (b) $u \cdot v = 0$ implies $u = 0$ or $v = 0$.

Proof. (a) If $u = \overline{(a, b)}$ and $0 = \overline{(x, x)}$, then
$$u \cdot 0 = \overline{(a, b)} \cdot \overline{(x, x)}$$
$$= \overline{(a \cdot x + b \cdot x, a \cdot x + b \cdot x)} \quad \text{definition of } \cdot$$
$$= 0 \quad \text{definition of } 0$$

(b) Suppose $u \cdot v = 0$. If $v = 0$, there is nothing to prove. If $v \neq 0$, we may write $0 \cdot v = 0$ by part (a) and the commutative law. Then
$$u \cdot v = 0 \cdot v$$
and by Theorem D-3, it follows that $u = 0$.

Theorem D-5. An integer u is positive if and only if $0 < u$.

Proof. An integer u has been defined to be positive if $u = {}^+n = \overline{(n + x, x)}$ for $n, x \in N$. It is immediate that
$$\overline{(x, x)} = 0 < \overline{(n + x, x)} = u$$
by definition of $<$, since
$$x + x < (n + x) + x$$
Conversely, if $0 < u = \overline{(a, b)}$, i.e.,
$$\overline{(x, x)} < \overline{(a, b)}$$
then
$$x + b < a + x$$
by definition of $<$. Hence, $b < a$. From this it follows that $b + n = a$ for some $n \in N$. Therefore, $u = \overline{(a, b)} = \overline{(n + b, b)} = {}^+n$, which shows that u is positive.

It is customary to identify the set N with the set P of positive integers. It is clear that these sets are not the same, but the identification of these two sets is justified

by the following considerations. Let us map each natural number n onto the corresponding positive integer ^+n. Call this mapping σ:

$$\sigma: n \longrightarrow {}^+n$$

As well as being 1–1 and onto, the mapping σ has the following properties:

(a) $(n + m)\sigma = n\sigma + m\sigma$
(b) $(n \cdot m)\sigma = (n\sigma) \cdot (m\sigma)$
(c) if $n < m$, then $n\sigma < m\sigma$
(d) if $n\sigma < m\sigma$, then $n < m$

The properties (a)–(d) show that any information concerning $+$, $\cdot$ or $<$ for either N or P may be turned into information about the other set. It is in this sense that the sets N and P are identified. Now we prove (a)–(d).

(a) By definition of σ, this is equivalent to $^+(n + m) = {}^+n + {}^+m$. We have $^+n = \overline{(n + a, a)}$, $^+m = \overline{(m + b, b)}$ for some $a, b \in N$. Then

$$\begin{aligned}
{}^+n + {}^+m &= \overline{(n + a, a)} + \overline{(m + b, b)} \\
&= \overline{((n + a) + (m + b), a + b)} \\
&= \overline{((n + m) + (a + b), a + b)} \\
&= {}^+(n + m) \qquad \text{by definition of } {}^+(n + m)
\end{aligned}$$

(b) This is proved by a method similar to part (a).

(c) By definition of σ, we wish to show that if $n < m$, then $^+n < {}^+m$. But if $n < m$, then $n + (a + b) < m + (a + b)$, which implies that

$$\overline{(n + a, a)} < \overline{(m + b, b)}$$

or

$$^+n < {}^+m$$

(d) The proof of part (d) is similar to the proof of part (c).

Summary of Properties of Z

The integers comprise a set Z with two binary operations, $+$ and $\cdot$, and an order relation, $<$, for which the following properties hold:

1. *Commutative Laws:* For all $u, v \in Z$

$$u + v = v + u \quad \text{and} \quad u \cdot v = v \cdot u$$

2. *Associative Laws:* For all $u, v, w \in Z$

$$(u + v) + w = u + (v + w) \text{ and } (u \cdot v) \cdot w = u \cdot (v \cdot w)$$

3. *Distributive Law:* For all $u, v, w \in Z$

$$u \cdot (v + w) = u \cdot v + u \cdot w$$

4. *The Identity Property of* 0: For all $u \in Z$

$$u + 0 = u$$

5. *The Identity Property of 1*: For all $u \in Z$
$$u \cdot 1 = u$$
6. *The Trichotomy Property:* For all $u, v \in Z$, one and only one of the following is true:
$$u < v \qquad u = v \qquad v < u$$
7. *The Cancellation Laws:* For all $u, v, w \in Z$
   (a) $u < v$ if and only if $u + w < v + w$
   (b) if $w > 0$, then $u < v$ if and only if $u \cdot w < v \cdot w$
   (c) if $w \neq 0$ and $u \cdot w = v \cdot w$, then $u = v$
   (d) if $w < 0$, then $u < v$ if and only if $v \cdot w < u \cdot w$

Bibliography

Alexandroff, P. S., *An Introduction to the Theory of Groups.* New York: Hafner Publishing Co., Inc., 1959.

Baumslag, B., and B. Chandler, *Theory and Problems of Group Theory.* New York: McGraw-Hill Book Co., 1968.

Grossman, Israel, and Wilhelm Magnus, *Groups and Their Graphs.* New York: Random House, Inc., 1964.

Ledermann, Walter, *Introduction to the Theory of Finite Groups.* New York: Interscience Publishers, Inc., 1957.

Lieber, Lillian, *Galois and the Theory of Groups.* New York: The Galois Institute of Mathematics and Art, 1932.

Macdonald, Ian D., *The Theory of Groups.* New York: Oxford University Press, Inc., 1968.

Papy, Georges, *Groups.* New York: St. Martin's Press, Inc., 1964.

Index

Abelian, 232
algorithm, 110

binary operations, 68, 70, 195
   commutative, 71
   associative, 71, 196

cartesian products, 33
class
   equivalence, 53
complement, 19
congruence
   modulo m, 121
   motions, 134
cosets, 184
   index of, 187
cycles
   disjoint, 176
   transposition, 176

direct products, 226
distributive law, 17, 83, 257, 262, 265, 272

division algorithm, 99

element, 159
   identity, 129, 265, 272, 273
   order of, 160

Fermat, theorem of, 186
Fundamental theorem of arithmetic, 116
functions, 57

Galois, theorem of, 196
greatest common divisor, 107
groups, 129
   cyclic, 159, 163
   order of, 178
   symmetric, 174

homomorphic image, 217
homomorphism, 216

identity element, 196, 209, 265, 272, 273
integer, 59, 263

integer (cont.)
  addition of, 264
  associative laws, 265, 272
  cancellation laws, 265, 271, 273
  commutative laws, 265, 272
  distributive law, 265, 272
  identity, 265, 272, 273
  inverse property, 265
  multiplication of, 264
  order relation on, 268
  trichotomy law, 269, 273
isomorphism, 205, 207

kernel, 217
Klein, Felix, 158
Klein Four-group, 158, 230

Lagrange, theorem of, 186
least common multiple, 107
least member, 84
linear combination, 110

mapping, 34, 57
  domain of, 34
  image of, 58
  inverse, 36, 37
  onto, 59
mathematical induction, 85
  principle of, 86
multiple, 100

natural number, 116
  associative laws, 262
  cancellation laws, 262
  commutative laws, 262
  distributive law, 262
  prime, 116
  product of, 116
  relatively prime, 117
  trichotomy property of, 260, 262
number theory, 83

operation, 9
ordered pair, 57

partition, 51
Peano axioms, 254
permutations, 171
  even, 177
  odd, 177

quaternion group, 237
quotient, 101
quotient group, 194, 197

relation, 34, 41
  equivalence, 41
  reflexive, 41, 42
  symmetric, 41, 43
  transitive, 41
remainder, 101
representative, 53
right regular reoresentation, 210
rotation, 132

set, 3
  associative laws, 16
  commutative laws, 16
  complement, 19
  distributive law, 17
  equality, 19
  idempotent law, 17
  intersection, 8, 17
  null, 4
  union, 8
  unit, 3
  singleton, 3
  subset of, 3
subgroup, 162
  cyclic, 196
  normal, 194
subset, 3

well-ordering property, 84, 261